Compact Textbooks in Mathematics

 Birkhäuser

Compact Textbooks in Mathematics

This textbook series presents concise introductions to current topics in mathematics and mainly addresses advanced undergraduates and master students. The concept is to offer small books covering subject matter equivalent to 2- or 3-hour lectures or seminars which are also suitable for self-study. The books provide students and teachers with new perspectives and novel approaches. They feature examples and exercises to illustrate key concepts and applications of the theoretical contents. The series also includes textbooks specifically speaking to the needs of students from other disciplines such as physics, computer science, engineering, life sciences, finance.

- **compact**: small books presenting the relevant knowledge
- **learning made easy**: examples and exercises illustrate the application of the contents
- **useful for lecturers**: each title can serve as basis and guideline for a semester course/lecture/seminar of 2–3 hours per week.

More information about this series at http://www.springer.com/series/11225

Jürgen Voigt

A Course on Topological Vector Spaces

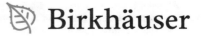

Jürgen Voigt
Institut für Analysis
Fakultät Mathematik
Technische Universität Dresden
Dresden, Germany

ISSN 2296-4568   ISSN 2296-455X   (electronic)
Compact Textbooks in Mathematics
ISBN 978-3-030-32944-0   ISBN 978-3-030-32945-7   (eBook)
https://doi.org/10.1007/978-3-030-32945-7

Mathematics Subject Classification (2010): 46A03, 46-01, 46A20, 46A08, 46A13

This book is published under the imprint Birkhäuser, www.birkhauser-science.com by the
registered company Springer Nature Switzerland AG.
The registered company address is: Gewerbestrasse 11, 6330 Cham, Switzerland

Preface

The theory of topological vector spaces – as a branch of functional analysis, motivated by applications and pushed forward also for abstract reasons – was developed over a long period of time, say, starting in the 1940s, and adopted in universities for teaching in the 1960s, when it became mandatory for advanced students in analysis to acquire knowledge in this topic. It was indeed in the late 1960s that I attended a course, given by the late Prof. W. Roelcke, University of Munich, on topological vector spaces and received my fundamental education in this area. When working in partial differential equations, operator theory, or some other topics in functional analysis, I always appreciated my knowledge in this abstract part of functional analysis, mainly as a somehow always present background.

It was in 2011, during discussions with some graduate students and young colleagues, that I discovered that they did not have, and missed, this kind of background – and in fact asked how they should have acquired it, due to the lack of offered courses. It was then that I decided to teach a course on this topic.

Due to the nature of the course, the book is certainly not intended to give an exhaustive treatment of the topic. The background the reader should have is the material presented in a basic course in functional analysis. In fact, the first two chapters of the book contain topics which mostly had been treated already in my basic course on functional analysis. Also, as seen immediately from the table of contents, the course is directed toward the theory of locally convex spaces.

The intended main objective of the course was the treatment of topologies for dual pairs, fundamental properties of which are contained in Chapters 3 to 6 – but unavoidably dual pairs are always present in the theory of locally convex spaces. In particular, the introduction of polar topologies and the Mackey–Arens theorem can be considered as a minimal kernel for the treatment of dual pairs. The topics of Chapters 8 to 11, reflexivity, completeness, locally convex final topology including applications, and compactness, are still pretty standard for the course.

Having covered these basic topics, I decided to present a choice of results which are of interest even for Banach spaces but need the theory of locally convex spaces. These are the Krein–Šmulian theorem in Chapter 12, the Eberlein–Šmulian theorem in Chapter 13 and the theorem of Krein in Chapter 14. Having talked so much about weakly compact sets in Chapters 13 and 14, I used Chapter 15 to present an important nontrivial example, in the form of weakly compact sets in L_1-spaces. Finally, in Chapter 16, I thought it of interest to present an example where it is possible to determine the bidual of a locally convex space for which one does not have an explicit representation of the dual.

The topics of Chapter 7, on Fréchet spaces, and Chapter 17, the Krein–Milman theorem, were not part of the actual course.

Clearly, in a presentation of results that have been developed over a long time, one cannot expect much originality. Nevertheless, desiring to proceed to interesting topics as fast as possible, I tried to present a streamlined approach, omitting many sidelines which might be interesting but not directly contributing to the aim I had in mind. According to my personal tastes in reading, I preferred a concise style where all needed ingredients are mentioned, but some active collaboration of the reader is required.

It may seem somewhat strange that I delegated the Hahn–Banach theorem and the uniform boundedness theorem to appendices. The reason is that I considered them as belonging to the prerequisites covered in a basic course on functional analysis (and in fact in the course itself, they were not included).

At the beginning of each chapter, I give a brief outline of topics treated therein. In the notes at the end of each chapter I try to mention the sources for the main results, sometimes adding further comments.

In an index of notation and an index, the reader can find the explanation of the symbols and of the terminology used in the text.

Finally, I want to add acknowledgements of various kinds. First of all, I want to thank the late Prof. Walter Roelcke for his introduction to the topic. Next, it is a pleasure to thank my colleagues and friends for many years from Munich times, Peter Dierolf and the late Susanne Dierolf, for many discussions and exchanges on various topics in the area as well as for the collaboration with Peter Dierolf. Finally, to come to more recent times, my thanks go to the late Prof. John Horváth for communication on the manuscript and for encouragement. It is a pleasure to thank Sascha Trostorff, Hendrik Vogt, and Marcus Waurick for many discussions on various topics in the book, and to Sascha Trostorff and Marcus Waurick for reading the first manuscript and discovering gaps, errors, and misprints. Also, I am much obliged to Dirk Werner for various comments on contents, examples, and misprints; in particular, it was his suggestion to include a chapter on Fréchet spaces, because of their importance in analysis.

And last but not least, along another line, I thank my wife Marianne for lifelong support and patience.

Dresden, Germany   Jürgen Voigt
August 2019

Contents

Initial Topology, Topological Vector Spaces, Weak Topology

The main objective of this chapter is to present the definition of topological vector spaces and to derive some fundamental properties. We will also introduce dual pairs of vector spaces and the weak topology. We start the chapter by briefly recalling concepts of topology and continuity, thereby also fixing notation.

Let X be a set, $\tau \subseteq \mathcal{P}(X)$ (the power set of X). Then τ is called a **topology**, and (X, τ) is called a **topological space**, if

for any $\mathcal{S} \subseteq \tau$ one has $\bigcup \mathcal{S} \in \tau$,
for any finite $\mathcal{F} \subseteq \tau$ one has $\bigcap \mathcal{F} \in \tau$.

(This definition is with the understanding that $\bigcup \varnothing = \varnothing$, $\bigcap \varnothing = X$, with the consequence that always $\varnothing, X \in \tau$.) Concerning notation, we could also write

$$\bigcup \mathcal{S} = \bigcup_{U \in \mathcal{S}} U, \quad \bigcap \mathcal{F} = \bigcap_{A \in \mathcal{F}} A.$$

If $\mathcal{S} = (U_\iota)_{\iota \in I}$ or $\mathcal{F} = (A_n)_{n \in N}$ are families of sets, with N finite, then one can also write

$$\bigcup \{U_\iota \,;\, \iota \in I\} = \bigcup_{\iota \in I} U_\iota, \quad \bigcap \{A_n \,;\, n \in N\} = \bigcap_{n \in N} A_n.$$

The sets $U \in \tau$ are called **open**, whereas a set $A \subseteq X$ is called **closed** if $X \setminus A$ is open. For a set $B \subseteq X$ we define

$\mathring{B} \ (= \operatorname{int} B) := \bigcup \{U \,;\, U \in \tau, \ U \subseteq B\}$, the **interior** of B (an open set),
$\overline{B} \ (= \operatorname{cl} B) := \bigcap \{A \,;\, A \supseteq B, \ A \text{ closed}\}$, the **closure** of B (a closed set).

© Springer Nature Switzerland AG 2020
J. Voigt, *A Course on Topological Vector Spaces*, Compact Textbooks in Mathematics,
https://doi.org/10.1007/978-3-030-32945-7_1

For $x \in X$, a set $U \subseteq X$ is called a **neighbourhood** of x if $x \in \overset{\circ}{U}$, and the collection

$$\mathcal{U}_x := \{U \subseteq X \,; \, U \text{ neighbourhood of } x\}$$

is called the **neighbourhood filter** of x. (Note that $U \cap V \in \mathcal{U}_x$ if $U, V \in \mathcal{U}_x$.) A **neighbourhood base** $\mathcal{B}$ of x is a collection $\mathcal{B} \subseteq \mathcal{U}_x$ with the property that the neighbourhood filter coincides with the collection of supersets of sets in $\mathcal{B}$. (Note that neighbourhoods need not be open sets.)

A topological space (X, τ) is called **Hausdorff** if for any $x, y \in X$, $x \neq y$, there exist neighbourhoods U of x, V of y such that $U \cap V = \varnothing$.

If (X, d) is a **semi-metric space**, i.e., X is a set and the **semi-metric** $d \colon X \times X \to [0, \infty)$ is symmetric and satisfies $d(x, x) = 0$ ($x \in X$) as well as the **triangle inequality**

$$d(x, y) \leqslant d(x, z) + d(z, y) \qquad (x, y, z \in X),$$

then d induces a topology τ_d on X: A set $U \subseteq X$ is defined to be open if for all $x \in U$ there exists $r > 0$ such that $B(x, r) \subseteq U$, where

$$B(x, r) = B_X(x, r) = B_d(x, r) := \{y \in X \,; \, d(y, x) < r\}$$

is the **open ball** with centre x and radius r. The corresponding **closed ball** will be denoted by

$$B[x, r] = B_X[x, r] = B_d[x, r] := \{y \in X \,; \, d(y, x) \leqslant r\}.$$

(We mention that our definition of 'semi-metric' often runs under the name 'pseudo-metric'; we found our notation more convenient, as it is parallel to 'semi-norm', mentioned later.) The topology τ_d is Hausdorff if and only if d is a **metric**, i.e., additionally to the previous properties one has that $d(x, y) = 0$ implies $x = y$.

A topological space (X, τ) is called **(semi-)metrisable** if there exists a (semi-)metric on X such that $\tau = \tau_d$.

If $\tau \supseteq \sigma$ are topologies on a set X, then τ is said to be **finer** (or **stronger**) than σ, and σ is said to be **coarser** (or **weaker**) than τ. The **trivial topology** $\{\varnothing, X\}$ is the coarsest topology on X, and the **discrete topology** $\mathcal{P}(X)$, i.e., the collection of all subsets of X, is the finest topology on X.

Let (X, τ), (Y, σ) be topological spaces, $f \colon X \to Y$, $x \in X$. Then f is **continuous at** x if $f^{-1}(V)$ is a neighbourhood of x, for all neighbourhoods V of $f(x)$. The mapping f is called **continuous**, if f is continuous at every $x \in X$, and this is equivalent to the property that $f^{-1}(V) \in \tau$ for all $V \in \sigma$. The mapping f is a **homeomorphism**, if f is continuous and bijective, and the inverse $f^{-1} \colon Y \to X$ is also continuous.

Remark 1.1 Let X be a set, $\Gamma \subseteq \mathcal{P}(\mathcal{P}(X))$ a set of topologies. Then it is easy to see that $\bigcap \Gamma$ is a topology on X. In order to spell this out more explicitly, we note that

$$\bigcap \Gamma = \bigcap_{\tau \in \Gamma} \tau = \{A \subseteq X; \ A \in \tau \text{ for all } \tau \in \Gamma\}.$$

(In this case, because of the subscript '$\tau \in \Gamma$', $\bigcap \tau$ does *not* mean $\bigcap_{U \in \tau} U$.) $\triangle$

Let X be a set, $\mathcal{S} \subseteq \mathcal{P}(X)$. Then

$$\text{top}\,\mathcal{S} := \bigcap \{\tau; \ \tau \text{ topology on } X, \ \tau \supseteq \mathcal{S}\}$$

is the coarsest topology containing $\mathcal{S}$, called the **topology generated by** $\mathcal{S}$, and $\mathcal{S}$ is called a **subbase** of top $\mathcal{S}$.

If τ is a topology, $\mathcal{B} \subseteq \tau$, and for all $U \in \tau$ one has that

$$U = \bigcup \{V \in \mathcal{B}; \ V \subseteq U\},$$

then $\mathcal{B}$ is called a **base** for τ. If $\mathcal{S}$ is a subbase of τ, then it is not difficult to show that

$$\mathcal{B} := \left\{\bigcap \mathcal{F}; \ \mathcal{F} \subseteq \mathcal{S}, \ \mathcal{F} \text{ finite}\right\} \tag{1.1}$$

is a base of τ.

Let X be a set. Let I be an index set (i.e., a set whose elements we use as indices), and for $\iota \in I$ let (X_ι, τ_ι) be a topological space and $f_\iota \colon X \to X_\iota$ a mapping. The topology

$$\text{top}\{f_\iota^{-1}(U_\iota); \ U_\iota \in \tau_\iota, \ \iota \in I\} \tag{1.2}$$

is the coarsest topology on X for which all mappings f_ι are continuous; it is called the **initial topology** with respect to the family $(f_\iota; \ \iota \in I)$. A base of the initial topology is given by

$$\left\{\bigcap_{\iota \in F} f_\iota^{-1}(U_\iota); \ F \subseteq I \text{ finite}, \ U_\iota \in \tau_\iota \ (\iota \in F)\right\}; \tag{1.3}$$

this is a consequence of (1.1) and (1.2).

The **product topology** on $\prod_{\iota \in I} X_\iota$ is the initial topology with respect to the family $(\text{pr}_\iota; \ \iota \in I)$ of the canonical projections. A base of the product topology is given by

$$\left\{\prod_{\iota \in F} U_\iota \times \prod_{\iota \in I \setminus F} X_\iota; \ F \subseteq I \text{ finite}, \ U_\iota \in \tau_\iota \ (\iota \in F)\right\}.$$

The following theorem is an important key result on initial topologies, which will be used repeatedly in this treatise.

Theorem 1.2
Let (Y, σ), (X, τ), (X_ι, τ_ι) $(\iota \in I)$ be topological spaces, $g: Y \to X$, $f_\iota: X \to X_\iota$ $(\iota \in I)$, τ the initial topology with respect to $(f_\iota \, ; \, \iota \in I)$. Let $y \in Y$. Then:
(a) *g is continuous at y if and only if $f_\iota \circ g$ is continuous at y $(\iota \in I)$.*
(b) *g is continuous if and only if $f_\iota \circ g$ is continuous $(\iota \in I)$.*
(c) *The initial topology on Y with respect to g is the same as the initial topology with respect to $(f_\iota \circ g \, ; \, \iota \in I)$.*

Proof
(a) The necessity is clear. In order to show the sufficiency, let U be a neighbourhood of $g(y)$. There exist a finite set $F \subseteq I$ and $U_\iota \in \tau_\iota$ $(\iota \in F)$ such that $\bigcap_{\iota \in F} f_\iota^{-1}(U_\iota) \subseteq U$ is a neighbourhood of $g(y)$. (Recall that these sets constitute a base of the initial topology.) Therefore, the set

$$g^{-1}\left(\bigcap_{\iota \in F} f_\iota^{-1}(U_\iota)\right) = \bigcap_{\iota \in F} g^{-1}\left(f_\iota^{-1}(U_\iota)\right) = \bigcap_{\iota \in F}(f_\iota \circ g)^{-1}(U_\iota)$$

is a neighbourhood of y, and is a subset of $g^{-1}(U)$.

(b) is a consequence of (a).

(c) is an immediate consequence of (b).   □

Next we define topological vector spaces and derive some basic properties.

Let E be a vector space over the field $\mathbb{K}$ (where $\mathbb{K} \in \{\mathbb{R}, \mathbb{C}\}$), and let τ be a topology on E. Then τ is called a **linear topology**, and (E, τ) is called a **topological vector space**, if the mappings

$$a: E \times E \to E, (x, y) \mapsto x + y,$$
$$m: \mathbb{K} \times E \to E, (\lambda, x) \mapsto \lambda x$$

are continuous.

In a topological vector space (E, τ) we will denote the neighbourhood filter of zero by $\mathcal{U}_0$ (or $\mathcal{U}_0(E)$, or $\mathcal{U}_0(\tau)$).

Examples 1.3
(a) A vector space E with the trivial topology $\tau = \{\varnothing, E\}$ is a topological vector space.

(b) A vector space $E \neq \{0\}$ with the discrete topology is not a topological vector space. Indeed, it is easy to see that the scalar multiplication m is not continuous.

(c) The scalars $\mathbb{R}$ and $\mathbb{C}$ are topological vector spaces.

(d) Normed and semi-normed spaces are topological vector spaces.   △

For more explanation on Example 1.3(d) we recall that a **semi-norm** p on a vector space E is a mapping $p\colon E \to [0, \infty)$ satisfying

$p(x + y) \leqslant p(x) + p(y)$ $(x, y \in E)$, the **triangle inequality**,
$p(\lambda x) = |\lambda| p(x)$ $(x \in E,\ \lambda \in \mathbb{K})$, i.e., p is **absolutely homogeneous**.

The semi-norm p gives rise to a semi-metric d on E, defined by $d(x, y) := p(x - y)$ $(x, y \in E)$. Then the inequalities $p((x + y) - (x_0 - y_0)) \leqslant p(x - x_0) + p(y - y_0)$ and $p(\lambda x - \lambda_0 x_0) \leqslant |\lambda| p(x - x_0) + |\lambda - \lambda_0| p(x_0)$ $(x, x_0, y, y_0 \in E,\ \lambda, \lambda_0 \in \mathbb{K})$ show the continuity of addition and scalar multiplication. The semi-metric d is a metric if and only if p is a **norm**, i.e., if additionally $p(x) = 0$ implies $x = 0$, for $x \in E$.

In the following theorem we collect some basic properties of topological vector spaces.

Theorem 1.4

Let (E, τ) be a topological vector space. Then:

(a) *For all $x \in E$ the mapping $a_x\colon E \to E,\ y \mapsto x + y$ is a homeomorphism. The topology τ is determined by a neighbourhood base of zero.*

(b) *For all $\lambda \in \mathbb{K} \setminus \{0\}$ the mapping $m_\lambda\colon E \to E,\ x \mapsto \lambda x$ is a homeomorphism.*

(c) *Each $U \in \mathcal{U}_0(E)$ is **absorbing**, i.e., for all $x \in E$ there exists $\alpha > 0$ such that $x \in \lambda U$ for all $\lambda \in \mathbb{K}$ with $|\lambda| \geqslant \alpha$.*

(d) *For all $U \in \mathcal{U}_0(E)$ there exists $V \in \mathcal{U}_0(E)$ such that $V + V \subseteq U$.*

Proof

(a) It is sufficient to show that the mapping a_x is continuous. It is a consequence of Theorem 1.2 (and the definition of the product topology on $E \times E$) that the mapping

$$j_x\colon E \to E \times E,\ y \mapsto (x, y)$$

is continuous. Then $a_x = a \circ j_x$ is continuous, because the addition a is continuous. The last statement is then obvious. (Note that the topology is determined if for each point in the space one knows a neighbourhood base.)

(b) Similarly to (a), we note that the mapping

$$j_\lambda\colon E \to \mathbb{K} \times E,\ x \mapsto (\lambda, x)$$

is continuous. Then the continuity of $m_\lambda = m \circ j_\lambda$ follows from the continuity of the scalar multiplication m.

(c) Similarly to part (a) one shows that the mapping $\mathbb{K} \ni \lambda \mapsto \lambda x \in E$ is continuous. Therefore there exists $\alpha > 0$ such that $\lambda x \in U$ for all $\lambda \in \mathbb{K}$ with $|\lambda| \leqslant \alpha$.

(d) Let $U \in \mathcal{U}_0(E)$. Then, by the continuity of the addition at the point $(0, 0)$, there exist $V_1, V_2 \in \mathcal{U}_0(E)$ such that $V_1 + V_2 \subseteq U$. Then $V := V_1 \cap V_2$ is as asserted. $\quad\square$

Next we introduce the concept of dual pairs of vector spaces, a central notion in our treatment.

A **dual pair** $\langle E, F \rangle$ consists of two vector spaces E, F over the same field $\mathbb{K}$ and a bilinear mapping $b = \langle \cdot, \cdot \rangle \colon E \times F \to \mathbb{K}$. The mapping b gives rise to mappings

$b_1 \colon E \to F^*$, defined by $b_1(x) := \langle x, \cdot \rangle$ $(x \in E)$,
$b_2 \colon F \to E^*$, defined by $b_2(y) := \langle \cdot, y \rangle$ $(y \in F)$,

where E^*, F^* denote the algebraic duals of E, F, respectively. The dual pair is **separating in E** if

$x \in E$, $\langle x, y \rangle = 0$ $(y \in F)$ implies that $x = 0$, i.e., b_1 is injective,

separating in F if

$y \in F$, $\langle x, y \rangle = 0$ $(x \in E)$ implies that $y = 0$, i.e., b_2 is injective,

and **separating**, if it is separating in E and F.

The **weak topology** $\sigma(E, F)$ on E with respect to the dual pair $\langle E, F \rangle$ is defined as the initial topology with respect to the family $(\langle \cdot, y \rangle; \ y \in F)$; the weak topology $\sigma(F, E)$ on F is defined analogously.

If $B \subseteq F$ is finite, then

$$U_B := \left\{ x \in E ; \ |\langle x, y \rangle| < 1 \ (y \in B) \right\}$$

is a $\sigma(E, F)$-neighbourhood of zero. A $\sigma(E, F)$-neighbourhood base of zero is given by

$$\left\{ U_B ; \ B \subseteq F \text{ finite} \right\};$$

see Remark 1.6.

The following theorem is basic for the theory and important for the construction of topological vector spaces; it shows (amongst other facts) that $\sigma(E, F)$ is a linear topology.

Theorem 1.5

Let E be a vector space, $\big((E_\iota, \tau_\iota); \ \iota \in I\big)$ a family of topological vector spaces, $f_\iota \colon E \to E_\iota$ linear maps $(\iota \in I)$, τ the initial topology on E with respect to $(f_\iota; \ \iota \in I)$. Then (E, τ) is a topological vector space.

Proof

First we show the continuity of the scalar multiplication $m \colon \mathbb{K} \times E \to E$. By Theorem 1.2 it is sufficient to show that $f_\iota \circ m \colon \mathbb{K} \times E \to E_\iota$ is continuous for all $\iota \in I$. For $\lambda \in \mathbb{K}, x \in E$, one has

$$f_\iota \circ m(\lambda, x) = f_\iota(\lambda x) = \lambda f_\iota(x) = m_\iota(\lambda, f_\iota(x)),$$

with m_ι denoting the scalar multiplication in E_ι; therefore $f_\iota \circ m = m_\iota \circ (\mathrm{id}_\mathbb{K} \times f_\iota)$. Noting that Theorem 1.2 implies that $\mathrm{id}_\mathbb{K} \times f_\iota \colon \mathbb{K} \times E \to \mathbb{K} \times E_\iota$ is continuous we obtain the assertion.

The continuity of the addition a in E is proved analogously: For $\iota \in I$, the continuity of $f_\iota \circ a = a_\iota \circ (f_\iota \times f_\iota)$ follows from the continuity of $f_\iota \times f_\iota \colon E \times E \to E_\iota \times E_\iota$ and the addition a_ι in E_ι. $\qquad\square$

Remark 1.6 If, in the situation of Theorem 1.5, $\mathcal{U}_\iota$ is a neighbourhood base of zero, for all $\iota \in I$, then a neighbourhood base of zero for the initial topology on E is given by

$$\left\{ \bigcap_{\iota \in F} f_\iota^{-1}(U_\iota); \ F \subseteq I \text{ finite}, \ U_\iota \in \mathcal{U}_\iota \ (\iota \in F) \right\}.$$

This follows from (1.3) $\qquad\qquad\qquad\qquad\qquad\qquad\qquad\qquad\qquad\qquad\qquad\qquad\triangle$

Examples 1.7
(a) The weak topologies $\sigma(E, F)$ and $\sigma(F, E)$, for a dual pair $\langle E, F \rangle$, are linear topologies.

(b) Let E be a vector space, P a set of semi-norms on E. Then the initial topology τ_P on E with respect to the mappings $\mathrm{id} \colon E \to (E, p)$ $(p \in P)$ is called the **topology generated by** P. Theorem 1.5 implies that τ_P is a linear topology.

(c) Let I be an index set. Then $\mathbb{K}^I$, with the product topology τ, the initial topology with respect to the projections $\mathrm{pr}_\kappa \colon \mathbb{K}^I \to \mathbb{K}$, $(x_\iota)_{\iota \in I} \mapsto x_\kappa$, is a topological vector space, by Theorem 1.5. With

$$c_\mathrm{c}(I) := \left\{ (y_\iota)_{\iota \in I} \in \mathbb{K}^I ; \ \{\iota \in I ; \ y_\iota \neq 0\} \text{ finite} \right\}$$

we form the dual pair $\langle \mathbb{K}^I, c_\mathrm{c}(I) \rangle$ by defining the duality bracket

$$\langle x, y \rangle := \sum_{\iota \in I} x_\iota y_\iota \qquad (x = (x_\iota)_{\iota \in I} \in \mathbb{K}^I, \ y = (y_\iota)_{\iota \in I} \in c_\mathrm{c}(I)).$$

Then $\tau = \sigma(\mathbb{K}^I, c_\mathrm{c}(I))$. Indeed, it is evident that $\tau \subseteq \sigma(\mathbb{K}^I, c_\mathrm{c}(I))$, because $\mathrm{pr}_\kappa x = \langle x, \delta_\kappa \rangle$, where $\delta_\kappa \in c_\mathrm{c}(I)$ is defined by $\delta_{\kappa\kappa} := 1$, $\delta_{\kappa\iota} := 0$ if $\iota \neq \kappa$. On the other hand, for each $y \in c_\mathrm{c}(I)$, the mapping $x \mapsto \langle x, y \rangle$ is a finite linear combination of canonical projections, hence continuous with respect to τ.

The product topology is also generated by the family of semi-norms $(p_\kappa)_{\kappa \in I}$, where $p_\kappa(x) := |x_\kappa|$ $(x = (x_\iota)_{\iota \in I} \in \mathbb{K}^I)$.

(d) Let X be a topological space, $E := C(X)$ the space of continuous functions $f \colon X \to \mathbb{K}$. For compact $K \subseteq X$ we define the semi-norm p_K, by

$$p_K(f) := \sup_{x \in K} |f(x)| \qquad (f \in C(X)),$$

and set

$$P := \{ p_K ; \; K \subseteq X \text{ compact} \}.$$

Then τ_P is the topology of **compact convergence**; it is a linear topology.

(e) Let $((E_\iota, \tau_\iota) ; \; \iota \in I)$ be a family of topological vector spaces, and let $E := \prod_{\iota \in I} E_\iota$. Then E, with the product topology, is a topological vector space. △

For a topological vector space (E, τ), the **dual**, or **dual space**, $(E, \tau)'$ is defined as the vector space of all continuous linear functionals on E. We will not always explicitly specify the topology of a topological vector space E, and accordingly, we will denote the dual of E by E' if it is clear from the context to which topology on E we refer.

By definition, every linear functional $\langle \cdot, y \rangle$, for $y \in F$, is continuous for $\sigma(E, F)$; the following result shows that the converse is also true.

Theorem 1.8
Let $\langle E, F \rangle$ be a dual pair. Let $\eta \in (E, \sigma(E, F))'$. Then there exists $y \in F$ such that $\eta(x) = \langle x, y \rangle$ $(x \in E)$. Expressed differently, one has $(E, \sigma(E, F))' = b_2(F)$.

For the proof we need a preparatory lemma from linear algebra.

Lemma 1.9 *Let E be a vector space, $\eta, \eta_1, \ldots, \eta_n \in E^*$,*

$$\bigcap_{j=1}^{n} \ker \eta_j \subseteq \ker \eta.$$

Then there exist $c_1, \ldots, c_n \in \mathbb{K}$ such that $\eta = \sum_{j=1}^{n} c_j \eta_j$.

Proof
(i) We start with a preliminary tool. Let F, G be vector spaces, $f \colon E \to F$ and $g \colon E \to G$ linear, g surjective, and $\ker g \subseteq \ker f$. Then there exists $\hat{f} \colon G \to F$ linear, such that $f = \hat{f} \circ g$.

In fact, $\hat{f}(g(x)) := f(x)$ $(x \in E)$ is well-defined: If $g(x) = g(x_1)$, then $x - x_1 \in \ker g \subseteq \ker f$, and therefore $f(x) = f(x_1)$. The linearity of $\hat{f}$ is then easy.

(ii) Apply (i) with $f = \eta$, $g = (\eta_1, \ldots, \eta_n) \colon E \to g(E) \subseteq \mathbb{K}^n$, to obtain $\hat{f} \colon g(E) \to \mathbb{K}$. There exists a linear extension $\hat{f} \colon \mathbb{K}^n \to \mathbb{K}$, and this extension is of the form

$$\hat{f}(y) = \sum_{j=1}^{n} c_j y_j \quad (y \in \mathbb{K}^n),$$

with suitable $(c_1, \ldots, c_n) \in \mathbb{K}^n$. Then $\eta = \hat{f} \circ (\eta_1, \ldots, \eta_n) = \sum_{j=1}^{n} c_j \eta_j$. □

Proof of Theorem 1.8

As η is continuous with respect to $\sigma(E, F)$, there exists a finite set $B \subseteq F$ such that

$$\eta(U_B) = \eta\big(\{x \in E \,;\; |\langle x, y \rangle| < 1 \ (y \in B)\}\big) \subseteq B_{\mathbb{K}}(0, 1)$$

(the open unit ball in $\mathbb{K}$), or expressed differently,

$$|\eta(x)| \leqslant \max_{y \in B} |\langle x, y \rangle| \quad (x \in E).$$

For $x \in E$ with $\langle x, y \rangle = 0$ $(y \in B)$ one concludes that $\eta(x) = 0$. From Lemma 1.9 we conclude that there exist $c_y \in \mathbb{K}$ $(y \in B)$ such that

$$\eta = \sum_{y \in B} c_y \langle \cdot, y \rangle = \langle \cdot, \sum_{y \in B} c_y y \rangle. \qquad \Box$$

Example 1.10

Coming back to $E = \mathbb{K}^I$ – see Example 1.7(c) – we note that Theorem 1.8 implies that $E' = \big(\mathbb{K}^I, \sigma(\mathbb{K}^I, c_\mathrm{c}(I))\big)' = c_\mathrm{c}(I)$. $\qquad \triangle$

From the definition it is clear that $\sigma(E, E')$ is the coarsest linear topology on E such that $E' \supseteq b_2(F)$, and Theorem 1.8 expresses that for this topology one even has $E' = b_2(F)$. Later we will also obtain a finest locally convex topology with this property; see Chapter 5.

Notes The material of the present chapter is standard, and it is rather impossible to give precise information where the contents originated. For the fundamental notions of topology we refer to [Bou07c]; in particular, our Theorem 1.2 is as in [Bou07c, Chap. 1, § 3, Proposition 4].

Concerning topological vector spaces and in particular locally convex spaces we include at this place a list of treatises on the subject, in principle in historical order: [Ban32], [Edw65], [Köt66], [Hor66], [Sch71] (first edition 1966), [Trè67], [Gro73], [RoRo73], [Rud91], [Wil78], [Bou07a] (new edition from 1981 of [Bou64a], [Bou64b]), [Jar81] [MeVo97], [Osb14], [BoSm17]. The beginning is marked by Banach's pioneering book. As mentioned in the preface, it was in the 1960s that the topic became "fashionable" also for teaching, and the treatises are of varying character, volume and focus. Wilansky's contribution is notable for its richness of exercises and examples, and we add Khaleelulla's Lecture Notes [Kha82] to the list as an abundant and well structured source of counterexamples.

The list indicated above contains only texts in which the main emphasis is on locally convex topological vector spaces. Many books on Banach space theory, functional analysis or operator theory contain also substantial parts on topological vector spaces. As examples, we mention the encyclopedic volume [DuSc58] and the treatises [Yos80], [Con90] and [Wer18].

Convexity, Separation Theorems, Locally Convex Spaces

Locally convex spaces are introduced as topological vector spaces possessing a neighbourhood base of zero consisting of convex sets. It is shown that then the topology can also be defined by a set of semi-norms. In order to show this and other features, we first treat separation properties. The final topic is the characterisation of (semi-)metrisability of locally convex spaces.

We recall that a **sublinear functional** on a vector space E is a mapping $p\colon E \to \mathbb{R}$ satisfying

$p(\lambda x) = \lambda p(x)$ $(\lambda \geqslant 0,\ x \in E)$, i.e., p is **positively homogeneous**,
$p(x + y) \leqslant p(x) + p(y)$ $(x, y \in E)$, i.e., p is **subadditive**.

We also recall that a set $A \subseteq E$ is called **convex** if $(1 - t)x + ty \in A$ for all $x, y \in A$, $t \in [0, 1]$.

Proposition 2.1 *Let E be a vector space. Then:*
(a) *If $p\colon E \to [0, \infty)$ is sublinear, then the sets*

$$A_p := \{x \in E;\ p(x) < 1\}, \qquad B_p := \{x \in E;\ p(x) \leqslant 1\}$$

are convex and absorbing.
(b) *If $A \subseteq E$ is convex and absorbing, then $p_A\colon E \to [0, \infty)$,*

$$p_A(x) := \inf\{\lambda \in (0, \infty);\ x \in \lambda A\} \quad (x \in E),$$

is sublinear, and

$$\{x \in E;\ p_A(x) < 1\} \subseteq A \subseteq \{x \in E;\ p_A(x) \leqslant 1\}.$$

*The mapping p_A is called the **Minkowski functional** (or **gauge**) of A.*

© Springer Nature Switzerland AG 2020
J. Voigt, *A Course on Topological Vector Spaces*, Compact Textbooks in Mathematics,
https://doi.org/10.1007/978-3-030-32945-7_2

Proof

(a) The convexity of A_p and B_p is an immediate consequence of the sublinearity of p. In order to show that A_p is absorbing, let $x \in E$. We will show below that there exists a constant $\alpha \geqslant 0$ such that $p(\gamma x) \leqslant \alpha$ for all $\gamma \in \mathbb{K}$ with $|\gamma| = 1$. Then, if $\lambda \in \mathbb{K}$, $|\lambda| > \alpha$, one obtains

$$p\left(\tfrac{1}{\lambda}x\right) = \tfrac{1}{|\lambda|}p\left(\tfrac{|\lambda|}{\lambda}x\right) < 1,$$

i.e., $x \in \lambda A_p$.

For $\mathbb{K} = \mathbb{R}$, the asserted inequality holds with $\alpha := \max\{p(x), p(-x)\}$. For $\mathbb{K} = \mathbb{C}$, it is straightforward to show that the inequality holds with $\alpha := \left(\max\{p(x), p(-x)\}^2 + \max\{p(ix), p(-ix)\}^2\right)^{1/2}$.

(b) It is easy to show that p_A is positively homogeneous. To verify the subadditivity, let $x, y \in E$, and assume that $p_A(x) + p_A(y) < 1$. Then there exist $\lambda, \mu > 0$ with $\lambda + \mu < 1$, $x \in \lambda A$, $y \in \mu A$. Then $x + y \in \lambda A + \mu A = (\lambda + \mu)A$, by the convexity of A, and therefore $p_A(x + y) < 1$.

If $x \in E$, $p_A(x) < 1$, then there exists $\lambda < 1$ with $x \in \lambda A \subseteq A$. If $x \in A$, then $p_A(x) \leqslant 1$ by definition. □

Theorem 2.2 (General separation theorem)

Let E be a topological vector space. Let $A, B \subseteq E$ be convex, non-empty, A open, $A \cap B = \varnothing$. Then there exists $x' \in E'$ such that

$$\operatorname{Re} x'(x) < \gamma := \inf_{y \in B} \operatorname{Re} x'(y) \tag{2.1}$$

for all $x \in A$.

This means that the 'affine real hyperplane' $\{x \in E\,;\ \operatorname{Re} x'(x) = \gamma\}$ 'separates' the sets A and B.

Lemma 2.3 *Let E be a topological vector space, $A \subseteq E$ open, $0 \neq x^* \in E^*$. Then $x^*(A)$ is open.*

Proof

There exists $x_0 \in E$ such that $x^*(x_0) > 0$. For $x \in A$ the set $A - x$ is absorbing (Theorem 1.4); hence, there exists $\varepsilon > 0$ such that $x + B_{\mathbb{K}}(0, \varepsilon)x_0 \subseteq A$. This implies

$$B_{\mathbb{K}}(x^*(x), \varepsilon x^*(x_0)) = x^*(x) + B_{\mathbb{K}}(0, \varepsilon)x^*(x_0) \subseteq x^*(A). \qquad \square$$

Proof of Theorem 2.2

It is sufficient to find $x' \in E' \setminus \{0\}$ such that (2.1) holds with '$\leqslant$'. From Lemma 2.3 one then obtains '$<$'. Without restriction we can assume that $\mathbb{K} = \mathbb{R}$; see Lemma A.1.

(i) Special case $B = \{x_0\}$: Without restriction $0 \in A$; otherwise choose $x_1 \in A$ and consider $A - x_1$, $x_0 - x_1$. Then A is absorbing. Let p_A be the Minkowski functional. Let $x'\colon \operatorname{lin}\{x_0\} \to \mathbb{R}$ be defined by $x'(x_0) := 1$. Then $x'(x_0) = 1 \leqslant p_A(x_0)$; therefore

$$x'(\lambda x_0) = \lambda \leqslant \lambda p_A(x_0) = p_A(\lambda x_0) \text{ for } \lambda \geqslant 0,$$
$$x'(\lambda x_0) = \lambda \leqslant 0 \leqslant p_A(\lambda x_0) \text{ for } \lambda \leqslant 0.$$

By the Hahn–Banach theorem, Theorem A.2, x' can be extended as $x' \in E^*$ such that $x'(x) \leqslant p_A(x)$ $(x \in E)$; in particular $x'(x) \leqslant 1 = x'(x_0)$ for $x \in A$.

It remains to show that x' is continuous, and for this it is sufficient to show continuity at 0 (because x' is linear and the topology is translation invariant; see Theorem 1.4(a)). For $\varepsilon > 0$, $x \in \varepsilon(A \cap (-A))$ one has $\pm\frac{1}{\varepsilon}x \in A$, therefore

$$\pm x'(x) = \varepsilon x'(\pm\tfrac{1}{\varepsilon}x) \leqslant \varepsilon p_A(\pm\tfrac{1}{\varepsilon}x) \leqslant \varepsilon.$$

(ii) General case: $A_1 := A - B = \bigcup_{y \in B}(A - y)$ is open, convex, $0 \notin A_1$. By part (i), there exists $x' \in E' \setminus \{0\}$ such that

$$x'(x - y) \leqslant x'(0) = 0, \text{ i.e., } x'(x) \leqslant x'(y)$$

for all $x \in A$, $y \in B$.   □

Example 2.4

In this example we illustrate that in a Hausdorff topological vector space E it can happen that the only convex open sets are the sets $\varnothing$, E (and this implies that $E' = \{0\}$).

Let $0 < p < 1$, and let (Ω, μ) be a measure space. Define

$$L_p(\mu) := \Big\{ f\colon \Omega \to \mathbb{K} \text{ measurable}\,;\ \int |f|^p\,d\mu < \infty \Big\}$$

(with functions a.e. equal identified; a vector space), with metric (!) d given by

$$d(f, g) := \int |f - g|^p\,d\mu.$$

It can be shown that $L_p(\mu)$ is a topological vector space, which is complete as a metric space.

We continue with the special $\Omega = (0, 1)$, with the Lebesgue measure λ. If $U \subseteq L_p(0, 1)$ is convex, open and non-empty, then we derive that $U = L_p(0, 1)$.

Let $\varepsilon > 0$. We show that $\operatorname{co} B(0, \varepsilon) = L_p(0, 1)$ (with 'co' denoting the convex hull and $B(0, \varepsilon)$ the ε-ball with centre 0 in $L_p(0, 1)$). Let $f \in L_p(0, 1)$. For $n \in \mathbb{N}$ there exist $f_1, \ldots, f_n \in L_p(0, 1)$ such that $f = f_1 + \cdots + f_n$, $\int |f_j|^p\,d\lambda = \frac{1}{n}\int |f|^p\,d\lambda$ $(j = 1, \ldots, n)$. (Subdivide the interval $(0, 1)$ suitably.) Then $f = \frac{1}{n}(nf_1 + \cdots + nf_n)$, and $\int |nf_j|^p\,d\lambda = n^p \int |f_j|^p\,d\lambda = n^{p-1}\int |f|^p\,d\lambda \to 0$ $(n \to \infty)$. Choosing n with $n^{p-1}\int |f|^p\,d\lambda < \varepsilon$ one concludes that $f \in \operatorname{co} B(0, \varepsilon)$.   △

For the separation of non-open convex sets by continuous linear functionals one needs an additional property of topological vector spaces.

A topological vector space is called **locally convex** if each neighbourhood of zero contains a convex neighbourhood of zero. In short, such a space will be called a **locally convex space**, and also the topology will be called **locally convex**.

Examples 2.5

(a) Semi-normed spaces are locally convex.

(b) If $(E_\iota;\ \iota \in I)$ is a family of locally convex spaces, E a vector space, $f_\iota\colon E \to E_\iota$ linear $(\iota \in I)$, then the initial topology on E is locally convex: If $\mathcal{U}_\iota$ is a neighbourhood base of zero of E_ι $(\iota \in I)$, then

$$\mathcal{U} := \left\{ \bigcap_{\iota \in F} f_\iota^{-1}(U_\iota);\ F \subseteq I \text{ finite},\ U_\iota \in \mathcal{U}_\iota\ (\iota \in F) \right\}$$

is a neighbourhood base of zero; see Remark 1.6.

By assumption, $\mathcal{U}_\iota$ can be chosen to consist of convex sets for all $\iota \in I$. Then $\mathcal{U}$ indicated above consists of convex sets.

(c) Let $\langle E, F \rangle$ be a dual pair. Then $\sigma(E, F)$ is locally convex.

(d) Let E be a vector space, P a set of semi-norms on E. Then the topology τ_P is locally convex. (In fact, the converse is also true, as will be shown below in Corollary 2.15.)   △

Theorem 2.6 (Separation theorem in locally convex spaces)
Let E be a locally convex space. Let $B \subseteq E$ be convex and closed, $x_0 \in E \setminus B$. Then there exists $x' \in E'$ such that

$$\operatorname{Re} x'(x_0) < \inf_{x \in B} \operatorname{Re} x'(x).$$

Lemma 2.7 *Let E be a topological vector space, $A \subseteq E$ convex. Then $\mathring{A}$ and $\overline{A}$ are convex.*

In the proof of this lemma we will need two technical details which we recall in the following remarks.

Remarks 2.8 (a) If X and Y are topological spaces, and $A \subseteq X$, $B \subseteq Y$, then $\overline{A \times B} = \overline{A} \times \overline{B}$. Indeed, the inclusion '$\subseteq$' holds because the set $\overline{A} \times \overline{B} = (\overline{A} \times Y) \cap (X \times \overline{B})$ is closed. On the other hand, if $(x, y) \in \overline{A} \times \overline{B}$, then for any neighbourhoods U of x, V of y the sets $U \cap A$, $V \cap B$ are non-empty, therefore $(U \times V) \cap (A \times B) \neq \varnothing$, and this implies that $(x, y) \in \overline{A \times B}$.

(b) Let X, Y be topological spaces, $f\colon X \to Y$ continuous, and $B \subseteq X$. Then $f(\overline{B}) \subseteq \overline{f(B)}$. Indeed, $f^{-1}(\overline{f(B)})$ is closed and contains B, hence also $\overline{B}$. Therefore $f(\overline{B}) \subseteq f(f^{-1}(\overline{f(B)})) \subseteq \overline{f(B)}$.   △

Proof of Lemma 2.7

(i) Let $x \in A$, $y \in \mathring{A}$, $0 < t \leqslant 1$. Then $(1 - t)x + ty \in (1 - t)x + t\mathring{A}$, and the last set is an open subset of A. Therefore $(1 - t)x + ty \in \mathring{A}$.

(ii) The mapping $f \colon \mathbb{R} \times E \times E \to E$, $(t, x, y) \mapsto (1 - t)x + ty$, is continuous, and $f([0, 1] \times B \times B) \subseteq B$ if and only if $B \subseteq E$ is convex. The properties mentioned in Remarks 2.8 imply

$$f([0, 1] \times \overline{A} \times \overline{A}) = f(\overline{[0, 1] \times A \times A}) \subseteq \overline{f([0, 1] \times A \times A)} \subseteq \overline{A};$$

hence $\overline{A}$ is convex. □

Proof of Theorem 2.6

As $x_0 \notin B$, there exists $U \in \mathcal{U}_0$ such that $(x_0 + U) \cap B = \varnothing$; without loss of generality one can take U convex and open (Lemma 2.7). From Theorem 2.2 we obtain the existence of $x' \in E'$ such that

$$\operatorname{Re} x'(x) < \inf_{y \in B} \operatorname{Re} x'(y)$$

for all $x \in x_0 + U$. □

Corollary 2.9 *Let E be a locally convex space.*

(a) *Let $E_0 \subseteq E$ be a closed subspace, $x_0 \in E \setminus E_0$. Then there exists $x' \in E'$ such that $x'|_{E_0} = 0$, $x'(x_0) \neq 0$.*

(b) *E is Hausdorff if and only if $\langle E, E' \rangle$ is separating in E.*

Proof

(a) We apply Theorem 2.6 with the closed convex set $B := E_0$. As $x'(x_0) \notin x'(E_0)$, we conclude that $x'(E_0) = \{0\}$.

(b) If E is Hausdorff, then $\{0\}$ is closed, and part (a) shows that $\langle E, E' \rangle$ is separating in E. On the other hand, if $\langle E, E' \rangle$ is separating in E and $x_0 \neq 0$, then there exists $x' \in E'$ with $x'(x_0) = 1$. Then the sets $U_0 := \{x \in E;\ \operatorname{Re} x'(x) < 1/2\}$ and $U_1 := \{x \in E;\ \operatorname{Re} x'(x) > 1/2\}$ are disjoint neighbourhoods of 0 and x_0, respectively, and this implies that E is Hausdorff. □

Corollary 2.10 *Let $\langle E, F \rangle$ be a dual pair which is separating in E, and let $G \subseteq F$ be a linear subspace. Then G is $\sigma(F, E)$-dense in F if and only if $\langle E, G \rangle$ is separating in E.*

Proof

By Corollary 2.9(a), G is dense in $(F, \sigma(F, E))$ if and only if the only functional in $(F, \sigma(F, E))'$ vanishing on G is the zero functional. Now, as $(F, \sigma(F, E))' = b_1(E)$, by Theorem 1.8, the latter holds if and only if $x \in E$, $\langle x, y \rangle = 0$ for all $y \in G$ implies $x = 0$, and this just means that $\langle E, G \rangle$ is separating in E. □

If E is a locally convex space, then the topology $\sigma(E, E')$ will be called the **weak topology** of E. The following result on the closure of convex sets is very important; it will often be used in the sequel.

Corollary 2.11   (Mazur) *Let E be a locally convex space, and let $B \subseteq E$ be a convex set. Then:*

(a)   *B is closed if and only if B is weakly closed.*

(b) $\overline{B} = \overline{B}^{\sigma(E,E')}$.

Proof

Let τ denote the topology of E.

(a) If B is weakly closed, then $\tau \supseteq \sigma(E, E')$ implies that B is closed. Now suppose that B is closed, and let $x_0 \in E \setminus B$. Theorem 2.6 yields a functional $x' \in E'$ such that

$$\operatorname{Re} x'(x_0) < \inf_{x \in B} \operatorname{Re} x'(x).$$

Therefore there exists $\varepsilon > 0$ such that $B(x'(x_0), \varepsilon) \cap \{x'(x);\ x \in B\} = \varnothing$, and thus $x'^{-1}(B(x'(x_0), \varepsilon)) \cap B = \varnothing$. As $x'^{-1}(B(x'(x_0), \varepsilon))$ is a $\sigma(E, E')$-neighbourhood of x_0, one concludes that $E \setminus B$ is weakly open, i.e., B is weakly closed.

(b) '$\subseteq$' follows from $\tau \supseteq \sigma(E, E')$. On the other hand, $\overline{B}$ is convex by Lemma 2.7, and hence therefore weakly closed by part (a).   □

As the next topic in this chapter we treat some fundamental issues concerning neighbourhoods of zero in topological vector spaces.

Let E be a vector space, $A \subseteq E$. The set A is called **balanced**, if $\lambda A \subseteq A$ for all $\lambda \in \mathbb{K}$ with $|\lambda| \leqslant 1$, and A is called **absolutely convex** if it is balanced and convex.

Remarks 2.12   (a) A set A is absolutely convex if and only if for all finite sets $B \subseteq A$, $\lambda_x \in \mathbb{K}$ $(x \in B)$ satisfying $\sum_{x \in B} |\lambda_x| \leqslant 1$ one has $\sum_{x \in B} \lambda_x x \in A$.

If $\mathcal{S} \subseteq \mathcal{P}(E)$, all $A \in \mathcal{S}$ balanced, then $\bigcap \mathcal{S}$ is balanced, and the same property holds with 'convex' and, 'absolutely convex' instead of 'balanced'.

For $A \subseteq E$, the **convex hull**

$$\operatorname{co} A := \bigcap \{B \subseteq E;\ A \subseteq B,\ B \text{ convex}\}$$

$$= \left\{ \sum_{x \in B} \lambda_x x;\ B \subseteq A \text{ finite},\ \lambda_x \in [0, 1]\ (x \in B),\ \sum_{x \in B} \lambda_x = 1 \right\}$$

is the smallest convex set containing A, the **balanced hull**

$$\operatorname{bal} A := \bigcap \{B \subseteq E;\ A \subseteq B,\ B \text{ balanced}\} = B_{\mathbb{K}}[0, 1] \cdot A$$

is the smallest balanced set containing A, and the **absolutely convex hull**

$$\text{aco } A := \bigcap \{B \subseteq E ; A \subseteq B, B \text{ absolutely convex}\}$$

$$= \left\{ \sum_{x \in B} \lambda_x x ; B \subseteq A \text{ finite}, \lambda_x \in B_{\mathbb{K}}[0, 1] (x \in B), \sum_{x \in B} |\lambda_x| \leq 1 \right\}$$

is the smallest absolutely convex set containing A.

(b) If A is a convex and absorbing set, then the Minkowski functional p_A is a semi-norm if and only if A is absolutely convex. $\triangle$

Lemma 2.13 *Let E be a topological vector space, $A \subseteq E$.*
(a) *Then $\bar{A} = \bigcap_{U \in \mathcal{U}_0} (A + U)$.*
(b) *Let A be balanced. Then $\bar{A}$ is balanced. If additionally $0 \in \mathring{A}$, then $\mathring{A}$ is balanced.*

Proof
(a) The element x belongs to $\bar{A}$ if and only if for all $U \in \mathcal{U}_0$ one has $(x - U) \cap A \neq \varnothing$, i.e., $x \in A + U$.

(b) If $x \in \bar{A}$, $|\lambda| \leq 1$, then $\lambda x \in \lambda \bar{A} = \overline{\lambda A} \subseteq \bar{A}$. If $x \in \mathring{A}$, $0 < |\lambda| \leq 1$, then $\lambda x \in \lambda \mathring{A} = \text{int}(\lambda A) \subseteq \mathring{A}$. $\square$

Theorem 2.14
Let E be a topological vector space.
(a) *Then the closed (resp., open) balanced neighbourhoods of zero constitute a neighbourhood base of zero.*
(b) *If E is locally convex, then the closed (resp., open) absolutely convex neighbourhoods constitute a neighbourhood base of zero.*

Proof
(a) Let $U \in \mathcal{U}_0$. There exists $U_1 \in \mathcal{U}_0$ such that $U_1 + U_1 \subseteq U$, therefore $\overline{U_1} \subseteq U$ (Lemma 2.13(a)). There exists $U_2 \in \mathcal{U}_0$ such that $\lambda U_2 \subseteq U_1$ whenever $|\lambda| < 1$. As a consequence,

$$V := \bigcup_{|\lambda| < 1} \lambda U_2 \subseteq U_1$$

is a balanced neighbourhood of zero, $\mathring{V} \subseteq \bar{V} \subseteq \overline{U_1} \subseteq U$, and $\mathring{V}$, $\bar{V}$ are balanced (Lemma 2.13(b)).

(b) is proved like part (a), but additionally one assumes that U_1 is convex, and one defines $V := \text{co} \bigcup_{|\lambda| < 1} \lambda U_2$, observes that the convex hull of a balanced set is balanced, and at the end also uses Lemma 2.7. $\square$

Corollary 2.15 *Let E be a locally convex space. Then there exists a set P of semi-norms, such that τ_P is the topology of E.*

Proof
If $\mathcal{U}$ is a neighbourhood base of zero consisting of absolutely convex sets, then the assertion follows with $P = \{p_U ; U \in \mathcal{U}\}$. $\qquad\square$

Corollary 2.16 *Let E be a locally convex space, $F \subseteq E$ a subspace, $y' \in F'$. Then there exists $x' \in E'$ such that $x'|_F = y'$.*

Proof
Let P be a set of semi-norms on E generating the topology of E, and assume without less of generality that P is 'directed' (defined below). Then there exist $p \in P$, $c \geqslant 0$ such that $|y'(y)| \leqslant cp(y)$ ($y \in F$). The Hahn–Banach theorem, Corollary A.3, implies that there exists a linear mapping $x' \colon E \to \mathbb{K}$ with $x'|_F = y'$ and $|x'(x)| \leqslant cp(x)$ ($x \in E$). The last inequality implies that $x' \in E'$. $\qquad\square$

An ordered set $(I, \leqslant)$ is called **directed** if for all $\iota_1, \iota_2 \in I$ there exists $\iota \in I$ such that $\iota_1 \leqslant \iota$ and $\iota_2 \leqslant \iota$.

We close this chapter with a characterisation of semi-metrisability of locally convex spaces.

Proposition 2.17 *Let E be a locally convex space. Then the following properties are equivalent:*
- (i) *E is semi-metrisable;*
- (ii) *there exists a countable neighbourhood base of zero;*
- (iii) *there exists a countable set P of semi-norms generating the topology of E;*
- (iv) *there exists a translation invariant semi-metric on E inducing the topology of E.*

Here, a semi-metric d is called **translation invariant** if $d(x, y) = d(x + z, y + z)$ for all $x, y, z \in E$. In the following lemma we single out part of the proof in a more general setup.

Lemma 2.18 *Let X be a set, let $\big((X_n, d_n)\big)_{n \in N}$, with $N \subseteq \mathbb{N}$, be a countable family of semi-metric spaces, and let $(f_n)_{n \in N}$ be a family of mappings $f_n \colon X \to X_n$. Then the initial topology on X with respect to the family $(f_n)_{n \in N}$ is semi-metrisable.*

Proof
By transporting the semi-metrics d_n to X,

$$\hat{d}_n(x, y) := d_n(f_n(x), f_n(y)) \quad (x, y \in X, \; n \in N)$$

(and then denoting $\hat{d}_n$ again by d_n) we see that we can transform the setup to the case where $(d_n)_{n \in N}$ is a family of semi-metrics on X.

It is easy to see that

$$d(x, y) := \sum_{n \in N} \min\{d_n(x, y), 2^{-n}\} \quad (x, y \in X) \tag{2.2}$$

defines a semi-metric on X. In order to show that this semi-metric induces the initial topology with respect to the family $(d_n)_{n \in N}$ it is therefore sufficient to show that they define the same neighbourhoods.

Let $x \in X$. For $n \in N$, $0 < \delta \leqslant 2^{-n}$, it is easy to check that $B_d(x, \delta) \subseteq B_{d_n}(x, \delta)$. This shows that each neighbourhood of x for the initial topology contains a d-ball with centre x.

On the other hand, for all $n \in N$ one checks that

$$\bigcap_{j \in N,\, 1 \leqslant j \leqslant n} B_{d_j}(x, 2^{-n}/n) \subseteq B_d(x, 2^{1-n}),$$

which implies that each d-ball with centre x contains a neighbourhood of x for the initial topology. □

Proof of Proposition 2.17

(i) $\Rightarrow$ (ii). If d is a semi-metric inducing the topology of E, then $\{B_d(0, 1/n); \, n \in \mathbb{N}\}$ is a countable neighbourhood base of zero.

'(ii) $\Rightarrow$ (iii)' is proved in the same way as Corollary 2.15.

(iii) $\Rightarrow$ (iv). Let (p_n) be a sequence of semi-norms inducing the topology of E. Then the semi-metrics d_n given by $d_n(x, y) := p_n(x - y)$ are translation invariant, hence the semi-metric d defined by (2.2) is also translation invariant, and d induces the topology of E.

'(iv) $\Rightarrow$ (i)' is trivial. □

Two semi-metrics d and e on a set X are said to be (topologically) **equivalent** if they induce the same topology. They are said to be **uniformly equivalent** if for each $\varepsilon > 0$ there exists $\delta > 0$ such that $B_d(x, \delta) \subseteq B_e(x, \varepsilon)$ as well as $B_e(x, \delta) \subseteq B_d(x, \varepsilon)$ for all $x \in X$. If E is a vector space, and d, e are equivalent translation invariant semi-metrics on E, then obviously they are uniformly equivalent.

A **Fréchet space** is a metrisable locally convex space which is complete with respect to a translation invariant metric. The previous comments imply that then the space is complete under each translation invariant metric. In the context of completeness for topological vector spaces, in Chapter 9, we will see that one can also describe a Fréchet space as a 'complete metrisable locally convex space', see Remark 9.1(f). Rather more surprisingly, a Fréchet space is also a 'completely metrisable locally convex space', i.e., the space is metrisable with a (not necessarily translation invariant!) metric, making it a complete metric space. For this result we refer to Theorem 7.11.

Examples 2.19

(a) Coming again back to $\mathbb{K}^I$ – see Examples 1.7(c) and 1.10 – and assuming that I is countably infinite, $I = \mathbb{N}$, we see that $\mathbb{K}^{\mathbb{N}}$ is a metrisable locally convex space. A translation invariant metric is given by

$$d(x, y) := \sum_{n \in \mathbb{N}} 2^{-n} \min\{|x_n - y_n|, 1\} \qquad (x, y \in \mathbb{K}^{\mathbb{N}}),$$

for instance. It is easy to see that $(\mathbb{K}^{\mathbb{N}}, d)$ is complete, hence $\mathbb{K}^{\mathbb{N}}$ is a Fréchet space.

(b) Let X be a Hausdorff locally compact topological space, and assume that there exists a sequence $(K_n)_{n \in \mathbb{N}}$ of compact sets with the property that every compact subset of X is contained in some K_n, i.e., X is **countable at infinity**.

Clearly, the topology of $C(X)$, defined in Example 1.7(d), is generated by the sequence $(p_{K_n})_{n \in \mathbb{N}}$ of semi-norms. A Cauchy sequence $(f_k)_{k \in \mathbb{N}}$ in $C(X)$, with respect to a translation invariant metric, is a Cauchy sequence with respect to all semi-norms p_{K_n}, hence a Cauchy sequence with respect to uniform convergence on the compact subsets of X. This implies that there exists $f \in C(X)$ such that $f_k \to f$ ($k \to \infty$) uniformly on the compact subsets of X. Hence $C(X)$ is a Fréchet space.

(c) The space s of **rapidly decreasing sequences** is defined by

$$s := \left\{x = (x_n) \in \mathbb{K}^{\mathbb{N}} ; \|x\|_k := \sup_n |x_n| n^k < \infty \text{ for all } k \in \mathbb{N}_0\right\},$$

with the increasing sequence of norms $(\|\cdot\|_k)_{k \in \mathbb{N}_0}$ generating the topology. It is easy to show that s is a Fréchet space.

To compute the dual space of s, we note that the topology of s is also generated by the sequence $(p_k)_{k \in \mathbb{N}_0}$ of norms, where

$$p_k(x) := \sum_{n=1}^{\infty} n^k |x_n| \qquad (x \in s, \ k \in \mathbb{N}_0).$$

Indeed, for all $x \in s$, $k \in \mathbb{N}_0$ one has

$$\|x\|_k \leqslant p_k(x) = \sum_{n=1}^{\infty} \frac{1}{n^2} n^{k+2} |x_n| \leqslant \left(\sum_{n=1}^{\infty} \frac{1}{n^2}\right) \|x\|_{k+2}.$$

Let $\eta \in s'$. Then there exist $C \geqslant 0$ and $k \in \mathbb{N}_0$ such that $|\eta(x)| \leqslant C p_k(x)$ ($x \in s$). Using the fact that $\ell_1' = \ell_\infty$ (and applying suitable isomorphisms between ℓ_1 and weighted ℓ_1-spaces and between ℓ_∞ and weighted ℓ_∞-spaces) we conclude that there exists $y = (y_n)_{n \in \mathbb{N}} \in \mathbb{K}^{\mathbb{N}}$ such that $\sup_n |y_n| n^{-k} \leqslant C$, i.e., $|y_n| \leqslant C n^k$ ($n \in \mathbb{N}$), and

$$\eta(x) = \sum_{n \in \mathbb{N}} x_n y_n =: \langle x, y \rangle \qquad (x \in s).$$

This shows that

$$t := s' = \bigcup_{k \in \mathbb{N}_0} \left\{ y \in \mathbb{K}^{\mathbb{N}} ; \; \sup_n |y_n| n^{-k} < \infty \right\}.$$

The sequences in t can be called 'polynomially bounded sequences', but in analogy to the dual of the Schwartz space – see Example 8.4(f) –, called the space of 'tempered distributions', we will call these sequences **tempered sequences**.

Later we will determine the strong topology $\beta(t, s)$ – see Example 7.7(a) –, verify that s is reflexive – see Example 8.4(a) –, and show that the strong topology on t is an inductive limit topology – see Example 10.7. △

Concerning Example 2.19(b), it is for simplicity that we assume the locally compact space to be Hausdorff. A topological space X is called σ-**compact** if there exists a sequence $(K_n)_{n \in \mathbb{N}}$ of compact subsets such that $X = \bigcup_{n \in \mathbb{N}} K_n$. It is not difficult to show that a Hausdorff locally compact space is countable at infinity if and only if it is σ-compact.

Notes As in Chapter 1, the material of the present chapter is standard. The separation theorems Theorem 2.2 and Theorem 2.6 can be found in many treatments, e.g. in [Edw65]. Corollary 2.11(a), for the case of normed spaces, is essentially due to Mazur [Maz33, Satz 3].

Some authors *define* locally convex spaces as spaces whose topology is defined by a set of semi-norms. The author prefers the definition that immediately reflects the name (and then to derive the above property for these spaces).

Polars, Bipolar Theorem, Polar Topologies

In a dual pair $\langle E, F \rangle$ one wants to define topologies on E associated with collections of suitable subsets of F. (This generalises the definition of the norm topology on the dual E' of a Banach space E, in this case for the dual pair $\langle E', E \rangle$.) Such a collection $\mathcal{M}$ defines a 'polar topology' on E, where the corresponding neighbourhoods of zero in E are polars of the members of $\mathcal{M}$. Examples of such topologies are the weak topology and the strong topology. In the first part of the chapter we define polars and investigate some of their properties.

Let $\langle E, F \rangle$ be a dual pair. For $A \subseteq E$ we define the **polar** (or 'absolute polar') $A^\circ \subseteq F$,

$$A^\circ := \left\{ y \in F ;\; |\langle x, y \rangle| \leqslant 1 \; (x \in A) \right\} = \left\{ y \in F ;\; \sup_{x \in A} |\langle x, y \rangle| \leqslant 1 \right\}.$$

Analogously, for $B \subseteq F$ we define $B^\circ := \left\{ x \in E ;\; |\langle x, y \rangle| \leqslant 1 \; (y \in B) \right\} \subseteq E$.

If E is a locally convex space and no space F is mentioned, then polars will be computed in the dual pair $\langle E, E' \rangle$.

If E is a vector space, we will use the symbol '$\bullet$' (instead of '$\circ$') to denote polars computed in the dual pair $\langle E, E^* \rangle$. (In fact, the symbol '$\bullet$' will also be used in some other situations, which will then be mentioned explicitly.)

Remark 3.1 There is no general consensus in the literature on how to define the polar. For instance, [Hor66] uses the definition as above, whereas in [Sch71] and [Wer18] the condition in the definition is 'Re $\langle x, y \rangle \leqslant 1$' instead of '$|\langle x, y \rangle| \leqslant 1$', and in [Bou07a] the corresponding condition is 'Re $\langle x, y \rangle \geqslant -1$'. $\triangle$

The following result expresses how to single out the continuous linear functionals within the larger set E^* of all linear functionals; it may serve as a first motivation for the notion of polars.

© Springer Nature Switzerland AG 2020
J. Voigt, *A Course on Topological Vector Spaces*, Compact Textbooks in Mathematics,
https://doi.org/10.1007/978-3-030-32945-7_3

Theorem 3.2

Let E be a topological vector space, $\mathcal{U}$ a neighbourhood base of zero of E. In the dual pair $\langle E, E^* \rangle$ one then has

$$E' = \bigcup_{U \in \mathcal{U}} U^{\bullet}.$$

Proof

The assertion is shown by the chain of equalities

$$E' = \left\{ x' \in E^* ; \text{ there exists } U \in \mathcal{U} \text{ such that } \sup_{x \in U} |\langle x, x' \rangle| \leqslant 1 \right\}$$

$$= \bigcup_{U \in \mathcal{U}} \left\{ x' \in E^* ; \sup_{x \in U} |\langle x, x' \rangle| \leqslant 1 \right\} = \bigcup_{U \in \mathcal{U}} U^{\bullet}. \qquad \square$$

Remarks 3.3 Let $\langle E, F \rangle$ be a dual pair, $A, B \subseteq E$. We note the following elementary properties of polars.

(a) If $A \subseteq B$, then $A^{\circ} \supseteq B^{\circ}$.

(b) If $\lambda \in \mathbb{K} \setminus \{0\}$, then $(\lambda A)^{\circ} = \frac{1}{\lambda} A^{\circ}$.

(c) If $\mathcal{A}$ is a collection of subsets of E, then $(\bigcup \mathcal{A})^{\circ} = \bigcap_{A \in \mathcal{A}} A^{\circ}$. △

Let E be a topological vector space. A set $A \subseteq E$ is called **bounded** if for all $U \in \mathcal{U}_0$ there exists $\lambda \in \mathbb{K}$ such that $A \subseteq \lambda U$, or equivalently, for all $U \in \mathcal{U}_0$ there exists $\alpha > 0$ such that $A \subseteq \lambda U$ for all $|\lambda| \geqslant \alpha$. (In a terminology introduced later, we could also say that A is bounded if it is absorbed by all neighbourhoods of zero.)

Lemma 3.4

(a) *Let E, F be topological vector spaces, $f : E \to F$ linear, continuous, $A \subseteq E$ bounded. Then $f(A)$ is bounded.*

(b) *Let E, E_ι $(\iota \in I)$ be topological vector space s, $f_\iota : E \to E_\iota$ linear, the topology on E the initial topology with respect to $(f_\iota; \iota \in I)$, and let $A \subseteq E$. Then A is bounded if and only if $f_\iota(A)$ is bounded for all $\iota \in I$.*

Proof

(a) Let $V \in \mathcal{U}_0(F)$. Then $f^{-1}(V) \in \mathcal{U}_0(E)$, and there exists $\lambda \in \mathbb{K}$ such that $A \subseteq \lambda f^{-1}(V)$. This implies that $f(A) \subseteq \lambda f(f^{-1}(V)) \subseteq \lambda V$.

(b) The necessity follows from (a). For the sufficiency, let $F \subseteq I$ be finite, and let $U_\iota \in \mathcal{U}_0(E_\iota)$ $(\iota \in F)$. Then there exists $\lambda > 0$ such that $f_\iota(A) \subseteq \lambda U_\iota$ $(\iota \in F)$; hence

$$A \subseteq f_\iota^{-1}(f_\iota(A)) \subseteq \lambda f_\iota^{-1}(U_\iota) \qquad (\iota \in F),$$

$$A \subseteq \lambda \bigcap_{\iota \in F} f_\iota^{-1}(U_\iota).$$

As the sets $\bigcap_{\iota \in F} f_\iota^{-1}(U_\iota)$ constitute a neighbourhood base of zero for the initial topology (see Remark 1.6), we obtain the assertion. $\qquad\square$

Proposition 3.5 *Let* $\langle E, F \rangle$ *be a dual pair,* $A \subseteq E$. *Then:*
(a) A° *is absolutely convex and* $\sigma(F, E)$-*closed.*
(b) A° *is absorbing if and only if* A *is* $\sigma(E, F)$-*bounded.*

Proof
(a) The representation

$$A^\circ = \bigcap_{x \in A} \langle x, \cdot \rangle^{-1} \big(B_{\mathbb{K}}[0, 1] \big)$$

shows that A° is an intersection of absolutely convex $\sigma(F, E)$-closed sets.
   (b) For $y \in F$, $\lambda \in \mathbb{K}$ one easily computes that

$$y \in \lambda A^\circ \iff \langle \cdot, y \rangle(A) \subseteq B_{\mathbb{K}}[0, |\lambda|].$$

This shows that A° is absorbing if and only if $\langle \cdot, y \rangle(A)$ is bounded for all $y \in F$, and by Lemma 3.4(b) the latter property is equivalent to the $\sigma(E, F)$-boundedness of A. $\qquad\square$

The following result plays a central role and will be used frequently.

Theorem 3.6 (Bipolar theorem)
Let $\langle E, F \rangle$ *be a dual pair,* $A \subseteq E$. *Then*

$$A^{\circ\circ} := (A^\circ)^\circ = \overline{\operatorname{aco} A}^{\,\sigma(E,F)}.$$

(Recall that the polar of the set $A^\circ \subseteq F$ *is a subset of* E.)

Proof
It is evident that $A \subseteq A^{\circ\circ}$; hence, $\overline{\operatorname{aco} A}^{\,\sigma(E,F)} \subseteq A^{\circ\circ}$, by Proposition 3.5(a).
   Let $x_0 \in E \setminus \overline{\operatorname{aco} A}^{\,\sigma(E,F)}$. By Theorem 2.6, and Theorem 1.8, there exists $y \in F$ such that

$$\sup_{x \in A} |\langle x, y \rangle| \leqslant \sup \big\{ \operatorname{Re}\langle x, y \rangle ;\; x \in \overline{\operatorname{aco} A}^{\,\sigma(E,F)} \big\} < \operatorname{Re}\langle x_0, y \rangle \leqslant |\langle x_0, y \rangle|;$$

by scaling y suitably we may assume that the left-hand side is equal to 1. Then $y \in A^\circ$, and from $1 < |\langle x_0, y \rangle|$ we conclude that $x_0 \notin A^{\circ\circ}$. $\qquad\square$

Remarks 3.7 (a) Let $(E, \| \cdot \|)$ be a normed space. We will use the "standard notation" $B_E := \{x \in E ; \|x\| \leqslant 1\}$ for the closed unit ball of E.

The bipolar theorem is a generalisation of Goldstine's theorem, asserting that $B_{E''} = \overline{B_E}^{\sigma(E'', E')}$. Indeed, in the dual pair $\langle E'', E' \rangle$ one has

$$B_{E''} = B_{E'}^{\circ} = (B_E^{\circ})^{\circ} = B_E^{\circ\circ} = \overline{B_E}^{\sigma(E'', E')},$$

where the first two equalities are a consequence of the definition of the norms in E' and E'', and where Theorem 3.6 is used in the last equality.

(b) Here is another version of the bipolar theorem (as stated in [MeVo97, Bipolar theorem 22.13]): Let E be a Hausdorff locally convex space, $A \subseteq E$ an absolutely convex set. Then $\overline{A} = A^{\circ\circ}$. (The polars act in the dual pair $\langle E, E' \rangle$.)

This is a consequence of Theorem 3.6, because $\overline{A} = \overline{A}^{\sigma(E, E')}$ by Corollary 2.11(b).   △

As a preliminary consideration for 'polar topologies' we note that for a dual pair $\langle E, F \rangle$ we want to define neighbourhoods of zero in E by polars B°. As these sets have to be absorbing, Proposition 3.5(b) implies that only $\sigma(F, E)$-bounded sets B are eligible for this procedure.

Let $\langle E, F \rangle$ be a dual pair,

$$\mathcal{B}_\sigma(F, E) := \{B \subseteq F ; \ B \ \sigma(F, E)\text{-bounded}\}.$$

For $B \in \mathcal{B}_\sigma(F, E)$ we define $q_B : E \to [0, \infty)$,

$$q_B(x) := \sup\{|\langle x, y \rangle| ; \ y \in B\} \quad (x \in E).$$

Then it is easy to see that q_B is a semi-norm, and

$$\{x \in E ; \ q_B(x) \leqslant 1\} = B^{\circ}.$$

Let $\mathcal{M} \subseteq \mathcal{B}_\sigma(F, E)$. Then the set $\{q_B ; \ B \in \mathcal{M}\}$ of semi-norms generates a locally convex topology $\tau_{\mathcal{M}}$ on E, the topology of uniform convergence on the sets of $\mathcal{M}$. The topologies defined in this way are called **polar topologies**. The topology generated by $\mathcal{M} = \mathcal{B}_\sigma(F, E)$ is called the **strong topology**, denoted by $\beta(E, F)$.

Correspondingly, one defines polar topologies on F.

Remarks 3.8 (a) $\tau_{\{\{y\} ; \ y \in F\}} = \sigma(E, F)$.

(b) If $\mathrm{lin} \bigcup \mathcal{M} = F$, then $\tau_{\mathcal{M}} \supseteq \sigma(E, F)$. Indeed, let $y \in B \in \mathcal{M}$. Then $q_{\{y\}} \leqslant q_B$, therefore id: $(E, \tau_{\mathcal{M}}) \to (E, q_{\{y\}})$ is continuous, and thus $\langle \cdot, y \rangle \in (E, \tau_{\mathcal{M}})'$. Since every $y \in F$ is a linear combination of elements from $\bigcup \mathcal{M}$, we conclude that $\langle \cdot, y \rangle \in (E, \tau_{\mathcal{M}})'$.

(c) It may be tempting to conjecture that the converse in (b) holds. However, this is not true, as the following example shows. Consider the dual pair $\langle \ell_1, \ell_\infty \rangle$, $B := \{x \in c_0 ; \ \|x\|_\infty \leqslant 1\}$, $\mathcal{M} := \{B\}$.

Then $B^\circ = B_{\ell_1}$ (the closed unit ball) and $\tau_{\mathcal{M}}$ is the norm topology on ℓ_1, which is finer than $\sigma(\ell_1, \ell_\infty)$. However, $\operatorname{lin} B = c_0 \neq \ell_\infty$.

(d) If $B \in \mathcal{B}_\sigma(F, E)$, then the semi-norm q_B is just the Minkowski functional of B°. Indeed, if $x \in E$, then $p_{B^\circ}(x) < 1$ if and only if there exists $\lambda \in (0, 1)$ such that $x \in \lambda B^\circ$, or equivalently, $|\langle x, y \rangle| \leqslant \lambda$ $(y \in B)$, and this holds if and only if $q_B(x) < 1$.

As a further observation we note that for $A, B \in \mathcal{B}_\sigma(F, E)$ one has $q_{A \cup B} = \max\{q_A, q_B\}$. △

In the following proposition, the notion 'directed' is used for a collection of sets, where the order refers to inclusion. Hence, '$\mathcal{M}$ directed' means that for any $A, B \in \mathcal{M}$ there exists $C \in \mathcal{M}$ such that $A \cup B \subseteq C$.

Proposition 3.9 *Let $\langle E, F \rangle$ be a dual pair, and let $\mathcal{M} \subseteq \mathcal{B}_\sigma(F, E)$ be directed.*
(a) *Then the collection*

$$\big\{\{x \in E;\ q_B(x) \leqslant \varepsilon\};\ B \in \mathcal{M},\ \varepsilon > 0\big\} = \big\{\varepsilon B^\circ;\ B \in \mathcal{M},\ \varepsilon > 0\big\}$$

constitutes a neighbourhood base of zero for $\tau_{\mathcal{M}}$.
(b) *Assume additionally that for all $A \in \mathcal{M}$, $\alpha > 0$ there exists $B \in \mathcal{M}$ such that $\alpha A \subseteq B$. Then the collection*

$$\{B^\circ;\ B \in \mathcal{M}\}$$

is neighbourhood base of zero for $\tau_{\mathcal{M}}$.

Proof
(a) The hypothesis implies that the set $P = \{q_B;\ B \in \mathcal{M}\}$ of semi-norms is directed, and therefore any finite intersection of closed ε-balls for semi-norms in P contains the closed ε-ball for some semi-norm in P.

(b) The additional hypothesis implies that any ε-ball for a semi-norm in P contains the closed 1-ball for some semi-norm in P, i.e., the polar of some $B \in \mathcal{M}$. □

Example 3.10
We show that for a Banach space E, the topologies $\beta(E, E')$ and $\beta(E', E)$ are the norm topologies.

(i) Let $B \subseteq E'$ be $\sigma(E', E)$-bounded, i.e., $\sup_{x' \in B} |x'(x)| < \infty$ for all $x \in E$. The uniform boundedness theorem implies that B is norm bounded, i.e., $B \subseteq c B_{E'}$ for some $c > 0$. Therefore $q_B \leqslant q_{c B_{E'}} = c q_{B_{E'}} = c \| \cdot \|_E$. Since $B_{E'}$ is bounded, we obtain the assertion for $\beta(E, E')$.

(ii) If $A \subseteq E$ is $\sigma(E, E')$-bounded, then it is also $\sigma(E'', E')$-bounded in E'', and therefore $\| \cdot \|_{E''}$-bounded by (i). Since $\| \cdot \|_E = \| \cdot \|_{E''}$ on E the assertion for $\beta(E', E)$ follows as in (i).

Whereas in this case always $(E, \beta(E, E'))' = E'$, the equality $(E', \beta(E', E))' = E$ holds if and only if E is reflexive. △

Let $\langle E, F \rangle$ be a separating dual pair. A locally convex topology τ on E is called **compatible with** $\langle E, F \rangle$ if $(E, \tau)' = F$.

For a locally convex space E, the **bidual** is defined as $E'' := (E', \beta(E', E))'$. The **canonical map** $\kappa : E \to E''$ is given by

$$\kappa(x)(x') := x'(x) \qquad (x' \in E', \ x \in E).$$

(It follows from $\beta(E', E) \supseteq \sigma(E', E)$ and Theorem 1.8 that the image of E under κ is contained in E''.) If E is Hausdorff, then κ is injective, and abbreviating one often writes $E \subseteq E''$, omitting the **canonical embedding** κ.

The space E is called **semi-reflexive** if E is Hausdorff and $E'' = E$ (as sets), and E is called **reflexive** if additionally the canonical embedding $\kappa : E \hookrightarrow (E'', \beta(E'', E'))$ is continuous. (In fact, in the latter case it follows that κ is an isomorphism; see the discussion at the beginning of Chapter 8.)

Notes The contents of this chapter are standard and basic for the investigation of locally convex topologies on dual pairs.

The Tikhonov and Alaoglu–Bourbaki Theorems

The central result of this chapter is the Alaoglu–Bourbaki theorem: Polars of neighbourhoods of zero in a locally convex space E are $\sigma(E', E)$-compact subsets of E'. As a consequence in a dual pair $\langle E, F \rangle$ one concludes that, for a locally convex topology τ on E with $(E, \tau)' = F$, one always has $\sigma(E, F) \subseteq \tau \subseteq \mu(E, F)$, where $\mu(E, F)$ is the Mackey topology on E, corresponding to the collection of absolutely convex $\sigma(F, E)$-compact subsets of F. As a prerequisite we show Tikhonov's theorem, and as a prerequisite to the proof of Tikhonov's theorem we introduce filters describing convergence and continuity of mappings in topological spaces.

Theorem 4.1 (Tikhonov)
Let $(X_\iota)_{\iota \in I}$ be a family of compact topological spaces. Then the product $\prod_{\iota \in I} X_\iota$ is compact.

We will prove this theorem here, even if it is rather part of general topology. However, the proof gives us the opportunity to introduce the notion of filters, which we will need anyway in the further treatment.

We recall that a topological space (X, τ) is called **compact** if every open covering of X (i.e., every collection $\mathcal{S} \subseteq \tau$ satisfying $\bigcup \mathcal{S} = X$) contains a finite subcovering (i.e., a finite collection $\mathcal{F} \subseteq \mathcal{S}$ such that $\bigcup \mathcal{F} = X$). Equivalently, X is compact if every collection $\mathcal{C}$ of closed subsets of X with the **finite intersection property** (i.e., $\bigcap \mathcal{F} \neq \varnothing$ for all finite $\mathcal{F} \subseteq \mathcal{C}$) satisfies $\bigcap \mathcal{C} \neq \varnothing$. Note that we use the notion of compactness in the sense that a compact space need not be Hausdorff.

A subset C of a topological space (X, τ) is called compact if $(C, \tau \cap C)$ is compact. (The topology $\tau \cap C := \{U \cap C \,;\, U \in \tau\}$ denotes the initial topology on C with respect to the injection $C \hookrightarrow X$, also called the **induced topology**.) If X is a Hausdorff topological

© Springer Nature Switzerland AG 2020
J. Voigt, *A Course on Topological Vector Spaces*, Compact Textbooks in Mathematics,
https://doi.org/10.1007/978-3-030-32945-7_4

space, and C is a compact subset, then it is easy to see that the complement of C is open, i.e., that C is closed.

Let X be a set. A **filter** $\mathcal{F}$ in X is a non-empty collection $\mathcal{F} \subseteq \mathcal{P}(X)$ satisfying the following properties:

$\varnothing \notin \mathcal{F}$;
if $A \in \mathcal{F}$, $A \subseteq B \subseteq X$, then $B \in \mathcal{F}$;
if $A, B \in \mathcal{F}$, then $A \cap B \in \mathcal{F}$.

A **filter base** $\mathcal{F}_0$ in X is a non-empty collection $\mathcal{F}_0 \subseteq \mathcal{P}(X)$ with:

$\varnothing \notin \mathcal{F}_0$;
if $A, B \in \mathcal{F}_0$, then there exists $C \in \mathcal{F}_0$ such that $C \subseteq A \cap B$.

If $\mathcal{F}_0$ is a filter base, then

$$\mathrm{fil}(\mathcal{F}_0) := \left\{ A \subseteq X ;\ \text{there exists } B \in \mathcal{F}_0 \text{ such that } B \subseteq A \right\}$$

is a filter, called the **filter generated by** $\mathcal{F}_0$. A filter $\mathcal{F}$ is called an **ultrafilter** if there is no filter properly containing $\mathcal{F}$.

Let now X be a topological space, $\mathcal{F}$ a filter in X, $x \in X$. Then $\mathcal{F}$ **converges to** x (or x **is a limit of** $\mathcal{F}$), $\mathcal{F} \to x$, if $\mathcal{U}_x \subseteq \mathcal{F}$. If $\mathcal{F}_0$ is a filter base, then one also writes $\mathcal{F}_0 \to x$ if the generated filter $\mathrm{fil}(\mathcal{F}_0)$ converges to x, i.e., if for all $U \in \mathcal{U}_x$ there exists $A \in \mathcal{F}_0$ with $A \subseteq U$. The point x is called a **cluster point** (also 'accumulation point') of a filter $\mathcal{F}$, if for all $U \in \mathcal{U}_x$, $A \in \mathcal{F}$ one has $U \cap A \neq \varnothing$, or equivalently, if $x \in \bigcap_{A \in \mathcal{F}} \bar{A}$.

Examples 4.2
Let X be a set.

(a) If $x \in X$, then $\mathcal{F}_0 := \{\{x\}\}$ is a filter base. The generated filter is called the **filter fixed at** x.

(b) If (x_n) is a sequence in X, then $\mathcal{F}_0 := \{\{x_j ;\ j \geqslant n\};\ n \in \mathbb{N}\}$ is a filter base. The generated filter is called an **elementary filter**.

If additionally X is a topological space and $x \in X$, then $\mathcal{F}_0 \to x$ if and only if $x_n \to x$ as $n \to \infty$.

(c) Let X be a topological space, $x \in X$. Then $\mathcal{U}_x$ is a filter (the neighbourhood filter of x). $\triangle$

Remarks 4.3 Let X be a set.

(a) If $\mathcal{F}$ is a filter in X, $A \subseteq X$ such that $A \cap B \neq \varnothing$ for all $B \in \mathcal{F}$, then obviously $\{A \cap B;\ B \in \mathcal{F}\}$ is a filter base, and the generated filter is **finer** than $\mathcal{F}$ (i.e., it contains $\mathcal{F}$).

(b) Let $\mathcal{F}$ be a filter. Then $\mathcal{F}$ is an ultrafilter if and only if for all $A \subseteq X$ one has $A \in \mathcal{F}$ or $X \setminus A \in \mathcal{F}$. (Necessity: If $A \cap B \neq \varnothing$ for all $B \in \mathcal{F}$, then (a) implies that there is a finer filter containing A, and this filter is equal to $\mathcal{F}$ because $\mathcal{F}$ is an ultrafilter; thus $A \in \mathcal{F}$. Otherwise there exists $B \in \mathcal{F}$ such that $A \cap B = \varnothing$, and then $X \setminus A \in \mathcal{F}$. Sufficiency: The condition implies that there is no finer filter.)

(c) For every filter $\mathcal{F}$ in X there exists a finer ultrafilter. This is an immediate consequence of Zorn's lemma. (In the proof that a maximal element is an ultrafilter one uses (a) and (b).)

(d) If X is a topological space, $\mathcal{F}$ is an ultrafilter in X, and $x \in X$ is a cluster point of $\mathcal{F}$, then $\mathcal{F} \to x$. (If $U \in \mathcal{U}_x$, then $U \cap A \neq \varnothing$ for all $A \in \mathcal{F}$, therefore $U \in \mathcal{F}$, because $\mathcal{F}$ is an ultrafilter.) $\triangle$

Remark 4.4 In our treatment we will use filters to discuss convergence and continuity in topological spaces. Filters generalise sequences – see Example 4.2(b) – which are sufficient for this purpose in metric spaces. (Another generalisation of sequences are 'nets', a notion that we will not need.) The proof of Theorem 4.1 becomes particularly nice with filters, but also for the discussion of completeness (Chapter 9) filters will be convenient. $\triangle$

Proposition 4.5 *Let X be a topological space. Then the following properties are equivalent:*
  (i) *X is compact;*
 (ii) *every filter in X possesses a cluster point;*
(iii) *every ultrafilter in X is convergent.*

Proof
(i) $\Rightarrow$ (ii). Let $\mathcal{F}$ be a filter in X. Then the collection $\{\overline{A}\,;\ A \in \mathcal{F}\}$ has the finite intersection property, and therefore $\bigcap_{A \in \mathcal{F}} \overline{A} \neq \varnothing$, i.e., $\mathcal{F}$ has a cluster point.

(ii) $\Rightarrow$ (i). Let $\mathcal{C} \subseteq \mathcal{P}(X)$ be a collection of closed sets with the finite intersection property. Then $\mathcal{F}_0 := \{\bigcap \mathcal{A}\,;\ \mathcal{A} \subseteq \mathcal{C} \text{ finite}\}$ is a filter base. The generated filter $\mathcal{F}$ has a cluster point, i.e., $\varnothing \neq \bigcap_{A \in \mathcal{F}} \overline{A} = \bigcap \mathcal{C}$.

'(ii) $\Rightarrow$ (iii)' is obvious, in view of Remark 4.3(d).

(iii) $\Rightarrow$ (ii). If $\mathcal{F}$ is a filter in X, then there exists a finer ultrafilter; see Remark 4.3(c). Every limit of this filter is a cluster point of $\mathcal{F}$. $\square$

Let X, Y be sets, $f\colon X \to Y$, $\mathcal{F}$ a filter in X. Then $f(\mathcal{F}) := \{f(A)\,;\ A \in \mathcal{F}\}$ is a filter base in Y, and the generated filter $\mathrm{fil}(f(\mathcal{F}))$ is called the **image filter**.

If $\mathcal{F}$ is an ultrafilter, then $f(\mathcal{F})$ is an ultrafilter base. Indeed, for $B \subseteq Y$ one has $f^{-1}(B) \in \mathcal{F}$ or $f^{-1}(Y \setminus B) \in \mathcal{F}$. In the first case one concludes that $f(f^{-1}(B)) \subseteq B \in \mathrm{fil}(f(\mathcal{F}))$, in the second case that $Y \setminus B \in \mathrm{fil}(f(\mathcal{F}))$.

Proposition 4.6
(a) *Let X, Y be topological spaces, $x \in X$, $\mathcal{F}$ a filter in X, $\mathcal{F} \to x$, $f\colon X \to Y$ continuous at x. Then $f(\mathcal{F}) \to f(x)$.*
(b) *Let X, X_ι ($\iota \in I$) be topological spaces, $f_\iota\colon X \to X_\iota$ ($\iota \in I$), and let the topology on X be the initial topology with respect to $(f_\iota)_{\iota \in I}$. Let $x \in X$, $\mathcal{F}$ a filter in X. Then $\mathcal{F} \to x$ if and only if $f_\iota(\mathcal{F}) \to f_\iota(x)$ for all $\iota \in I$.*

Proof

(a) Let V be a neighbourhood of $f(x)$. Then $f^{-1}(V)$ is a neighbourhood of x, and therefore $f^{-1}(V) \in \mathcal{F}$. From $f(f^{-1}(V)) \subseteq V$ one then obtains $V \in \mathrm{fil}(f(\mathcal{F}))$.

(b) The necessity is clear from (a). For the sufficiency let $U \in \mathcal{U}_x$. Then there exist a finite set $F \subseteq I$ and neighbourhoods U_ι of $f_\iota(x)$ ($\iota \in I$) such that $\bigcap_{\iota \in F} f_\iota^{-1}(U_\iota) \subseteq U$. There exists $A \in \mathcal{F}$ such that $f_\iota(A) \subseteq U_\iota$ ($\iota \in F$). Therefore

$$A \subseteq f_\iota^{-1}(f_\iota(A)) \subseteq f_\iota^{-1}(U_\iota) \quad (\iota \in F),$$

$$A \subseteq \bigcap_{\iota \in F} f_\iota^{-1}(U_\iota) \subseteq U.$$

$\square$

Proof of Theorem 4.1

Without restriction all $X_\iota \neq \varnothing$. Let $\mathcal{F}$ be an ultrafilter in $\prod_{\iota \in I} X_\iota$. Then $\mathrm{pr}_\iota(\mathcal{F})$ is an ultrafilter base in X_ι, therefore convergent by Proposition 4.5, $\mathrm{pr}_\iota(\mathcal{F}) \to x_\iota \in X_\iota$ ($\iota \in I$). Then Proposition 4.6(b) implies that $\mathcal{F} \to (x_\iota)_{\iota \in I}$.

$\square$

As in the case of Banach spaces Tikhonov's theorem implies the Banach–Alaoglu theorem, i.e., the closed dual ball is weak*-compact, we now derive the corresponding result for locally convex spaces.

Theorem 4.7 (Alaoglu–Bourbaki)
Let E be a locally convex space, $U \subseteq E$ a neighbourhood of zero. Then $U^\circ \subseteq E'$ is $\sigma(E', E)$-compact.

Lemma 4.8 *Let E be a vector space. Then E^* is closed in $\mathbb{K}^E$ with respect to the product topology.*

Proof

For $\lambda \in \mathbb{K}$, $x, y \in E$ the mapping

$$\varphi_{\lambda, x, y} \colon \mathbb{K}^E \to \mathbb{K}, \ f \mapsto f(\lambda x + y) - \lambda f(x) - f(y)$$

is continuous. (Note that, for $x \in E$, the mapping $\mathbb{K}^E \ni f \mapsto f(x) \in \mathbb{K}$ is one of the projections defining the product topology.) Therefore $E^* = \bigcap_{\lambda \in \mathbb{K}, x, y \in E} \varphi_{\lambda, x, y}^{-1}(0)$ is closed. $\square$

Proof of Theorem 4.7

Without loss of generality we may assume that U is absolutely convex. We note that $x' \in U^\circ$ if and only if $x' \in E^*$ and $|\langle x, x' \rangle| \leqslant p_U(x)$ ($x \in E$). The condition is clearly sufficient. On the other hand, if $x' \in U^\circ$, $x \in E$, $\lambda > p_U(x)$, then $\frac{1}{\lambda} x \in U$, $|\langle \frac{1}{\lambda} x, x' \rangle| \leqslant 1$, $|\langle x, x' \rangle| \leqslant \lambda$;

therefore, $|\langle x, x' \rangle| \leqslant p_U(x)$. This implies that

$$
\begin{aligned}
U^\circ &= \left\{ x' \in E^*; \; |\langle x, x' \rangle| \leqslant p_U(x) \; (x \in E) \right\} \\
&= \left\{ f \in \mathbb{K}^E; \; |f(x)| \leqslant p_U(x) \; (x \in E) \right\} \cap E^* \\
&= \left(\prod_{x \in E} B_{\mathbb{K}}[0, p_U(x)] \right) \cap E^*.
\end{aligned}
$$

Theorem 1.2 implies that the weak topology on E' and the product topology on $\prod_{x \in E} B_{\mathbb{K}}[0, p_U(x)]$ are the restrictions of the product topology on $\mathbb{K}^E = \prod_{x \in E} \mathbb{K}$ to these sets. Because of Lemma 4.8 it therefore follows that U° is a closed subset of the compact set $\prod_{x \in E} B_{\mathbb{K}}[0, p_U(x)]$. □

Let $\langle E, F \rangle$ be a dual pair. Let

$$
\mathcal{M}_\mu := \left\{ B \subseteq F; \; B \text{ absolutely convex and } \sigma(F, E)\text{-compact} \right\}.
$$

Obviously one has $\mathcal{M}_\mu \subseteq \mathcal{B}_\sigma(F, E)$. Then the polar topology

$$
\mu(E, F) := \tau_{\mathcal{M}_\mu}
$$

on E is called the **Mackey topology**. The Mackey topology $\mu(F, E)$ on F is defined correspondingly.

In the following Chapter 5 we will show that $(E, \mu(E, F))' = b_2(F)$, and that $\mu(E, F)$ is the strongest topology with dual $b_2(F)$, in the following sense: If $\langle E, F \rangle$ is a separating dual pair, then a locally convex topology τ on E is compatible with $\langle E, F \rangle$ if and only if $\sigma(E, F) \subseteq \tau \subseteq \mu(E, F)$.

In the last statement, the necessity of the condition is easily obtained from our treatment presented so far. If τ is compatible, the property $\sigma(E, F) \subseteq \tau$ follows from the definition of the topology $\sigma(E, F)$ (and Theorem 1.2), whereas the property $\tau \subseteq \mu(E, F)$ is a consequence of Theorem 4.7, as follows. The space (E, τ) possesses a neighbourhood base of zero $\mathcal{U}$ consisting of closed absolutely convex sets; hence

$$
\mathcal{M} := \left\{ U^\circ; \; U \in \mathcal{U} \right\} \subseteq \mathcal{M}_\mu,
$$

by Theorem 4.7, and therefore $\tau = \tau_{\mathcal{M}} \subseteq \tau_{\mathcal{M}_\mu} = \mu(E, F)$.

The definition of $\mathcal{M}_\mu$ suggests the question whether in a locally convex space the closed absolutely convex hull of a compact set is again compact. Example 4.10 given below shows that this is not always the case. We will show in Chapter 11 that it is true if E is quasi-complete (Corollary 11.5). In particular it is true if E is a Banach space ('Mazur's theorem'). In Chapter 14 we will show that it is also true for the weak topology in a Banach space ('Krein's theorem'). However, it is always true that the closed absolutely convex hull of a compact *convex* set is compact; this is the content of the following lemma. As a consequence one obtains $\mu(E, F) = \tau_{\mathcal{M}'_\mu}$ also for

$$\mathcal{M}'_\mu := \{B \subseteq F;\ B \text{ convex and } \sigma(F, E)\text{-compact}\}.$$

Lemma 4.9 *Let E be a topological vector space, and let $A \subseteq E$ be a compact convex subset. Then $\overline{\text{aco}}\, A$ is compact.*

Proof
(i) If $B \subseteq E$ is a balanced subset, then $\text{aco}\, B = \text{co}\, B$. This holds because

$$\text{co}\, B = \left\{\sum_{j=1}^{n} \lambda_j x_j\,;\ \lambda_1, \ldots, \lambda_n \in [0, 1],\ \sum_{j=1}^{n} \lambda_j = 1,\ x_1, \ldots, x_n \in B,\ n \in \mathbb{N}\right\}$$

is easily seen to be balanced.

(ii) If $\mathbb{K} = \mathbb{R}$, then $\text{bal}\, A = [-1, 1] \cdot A \subseteq \text{co}(A \cup (-A))$, and the latter set is compact (as the image of the compact set $\{(\lambda_1, \lambda_2) \in [0, 1]^2;\ \lambda_1 + \lambda_2 = 1\} \times A \times (-A)$ under the continuous mapping $(\lambda_1, \lambda_2, x_1, x_2) \mapsto \lambda_1 x_1 + \lambda_2 x_2$). Hence $\overline{\text{aco}}\, A = \overline{\text{co}(\text{bal}\, A)} \subseteq \text{co}(A \cup (-A))$ is compact.

(iii) If $\mathbb{K} = \mathbb{C}$, then

$$\text{bal}\, A = B_{\mathbb{C}}[0, 1] \cdot A \subseteq \sqrt{2}\,\text{co}\big(A \cup (iA) \cup (-A) \cup (-iA)\big),$$

where again the latter set is compact. The remaining argument is as in (ii). □

Example 4.10 (cf. [Kha82, Chap. II, Example 10])
Consider the dual pair $\langle c_c, \ell_1 \rangle$, where $c_c := \text{lin}\{e_n;\ n \in \mathbb{N}\}$, with the 'unit vectors' e_n in c_0 (or ℓ_1). The sequence $(2^n e_n)_n$ in ℓ_1 is $\sigma(\ell_1, c_c)$-convergent to 0; therefore $B := \{2^n e_n;\ n \in \mathbb{N}\} \cup \{0\}$ is $\sigma(\ell_1, c_c)$-compact. For $n \in \mathbb{N}$, the element $y^n := \sum_{j=1}^{n} e_j = \sum_{j=1}^{n} 2^{-j}(2^j e_j)$ belongs to $\text{co}\, B$. For a $\sigma(\ell_1, c_c)$-cluster point $y = (y_j)$ of the sequence (y^n), the coordinate y_j would have to be a cluster point of the sequence $(y_j^n)_n$, i.e., $y_j = 1$ ($j \in \mathbb{N}$). However, the element $(1, 1, 1, \ldots)$ does not belong to ℓ_1. This shows that the sequence $(y^n)_n$ does not have a cluster point, and therefore the set $\text{co}\, B$ is not relatively compact with respect to $\sigma(\ell_1, c_c)$. △

We include an additional information on metrisability in the context of the Alaoglu–Bourbaki theorem.

Proposition 4.11 *Let E be a separable locally convex space, $U \subseteq E$ a neighbourhood of zero. Then the topology $\sigma(E', E)$ is metrisable on $U^\circ \subseteq E'$.*

Proof
Let $A \subseteq E$ be a countable dense set, and denote by ρ the initial topology on E' with respect to the family $\big(E' \ni x' \mapsto \langle x, x' \rangle \in \mathbb{K}\big)_{x \in A}$. Then ρ is coarser than $\sigma(E', E)$, and ρ is metrisable, by Proposition 2.17 (where the denseness of A in E implies that ρ is Hausdorff).

As $(U^\circ, \sigma(E', E) \cap U^\circ)$ is compact by the Alaoglu–Bourbaki theorem, one concludes from Lemma 4.12, proved below, that $\rho \cap U^\circ = \sigma(E', E) \cap U^\circ$. □

For completeness we recall (from general topology) the following important basic observation concerning compactness.

Lemma 4.12 *Let* X, Y *be topological spaces,* X *compact,* Y *Hausdorff,* $f : X \to Y$ *continuous and bijective. Then* f *is a homeomorphism.*

Proof
We only have to show that f is an open mapping. Let $U \subseteq X$ be an open set. Then $X \setminus U$ is closed, hence compact. This implies that $Y \setminus f(U) = f(X \setminus U)$ is compact, hence closed, i.e., $f(U)$ is open. □

Notes Tikhonov's theorem is one of the basic theorems of topology, in some sense the first result in the development of set theoretic topology which does not come along with a straightforward 'evident' proof. Tikhonov (in early German transcription "Tychonoff") proved the theorem for compact intervals in [Tyc30] and mentioned later that the proof carries over to the general case. The main result of this chapter, the Alaoglu–Bourbaki theorem (Theorem 4.7), uses Tikhonov's theorem. For the case of normed spaces it usually is called the Banach–Alaoglu theorem, proved for the separable case by Banach [Ban32, VIII, § 5, Théorème 3] and for the general case by Bourbaki [Bou38, Corollaire de Théorème 1] (and shortly after by Alaoglu [Ala40, Theorem 1:3]). The first appearance of the general case may be in a paper of Arens [Are47, proof of Theorem 2]. (It is also contained in Bourbaki [Bou64b, Chap. IV, § 2.2, Proposition 2].) The Mackey topology was first defined and used by Arens [Are47]; we use the notation $\mu(E, F)$, for a dual pair $\langle E, F \rangle$, thereby following Floret [Flo80]. (A more traditional notation, used by many authors, would be $\tau(E, F)$, and the author has been told the reason for this notation: $\sigma(E, F)$ is the 'beginning' of the scale of compatible locally convex topologies, and $\tau(E, F)$ is the 'end'; like one often uses $[a, b]$ for intervals in $\mathbb{R}$, the idea is to use the neighbouring letters σ and τ in the Greek alphabet as the ends of the 'interval'. As we use 'τ' quite generally for topologies, we prefer Floret's notation. Anyway, 'σ' in weak topologies probably comes from the 's' in the German "schwach". The earliest place where the author could localise the use of '$\sigma(E, E')$' for the weak topology, is the note [Die40].)
Summarising the previous discussion, if the names given to theorems should indicate their authors, then the Banach–Alaoglu theorem should be called 'Banach–Bourbaki theorem', the Alaoglu–Bourbaki theorem should be called 'Bourbaki–Arens theorem', and the Mackey topology should be called 'Arens topology' (although in the latter case 'Arens–Mackey topology' would be equally justified).
Concerning Lemma 4.9, we refer to [Edw65, Remark 8.13.4(3)].

5

The Mackey–Arens Theorem

The first objective is to complete the discussion concerning compatible topologies on E for a dual pair $\langle E, F \rangle$, by showing that $(E, \mu(E, F))' = F$. In examples we discuss 'compatibility' for non-locally convex topologies. For the dual pair $\langle \ell_\infty, \ell_1 \rangle$ we compute the Mackey topology on ℓ_∞; in this treatment there will come up interesting properties of ℓ_1 which will turn out of importance in Chapter 15.

Theorem 5.1 (Mackey–Arens)
Let $\langle E, F \rangle$ be a dual pair.
(a) *Then $(E, \mu(E, F))' = b_2(F)$ (where $\mu(E, F)$ is the Mackey topology defined in the previous chapter).*
(b) *Assume additionally that $\langle E, F \rangle$ is separating. Then a locally convex topology τ on E is compatible with $\langle E, F \rangle$ if and only if $\sigma(E, F) \subseteq \tau \subseteq \mu(E, F)$.*

For the proof of the theorem we need further preparations.

Lemma 5.2 *Let E be a topological vector space, $A, B \subseteq E$ compact. Then $A + B$ is compact.*

Proof
Since the addition $a \colon E \times E \to E$ is continuous, the sum $A + B = a(A \times B)$ is compact. $\qquad\qquad\square$

In the following lemma as well as in the proof of Theorem 5.1, the symbol '$\circ$' will refer to polars taken in the dual pair $\langle E, F \rangle$, whereas '$\bullet$' refers to polars taken in $\langle E, E^* \rangle$. We also recall the mapping $b_2 \colon F \to E^*$, $b_2(y) := \langle \cdot, y \rangle$ ($y \in F$) (from Chapter 1).

© Springer Nature Switzerland AG 2020
J. Voigt, *A Course on Topological Vector Spaces*, Compact Textbooks in Mathematics,
https://doi.org/10.1007/978-3-030-32945-7_5

Lemma 5.3 *Let $\langle E, F \rangle$ be a dual pair, $B \subseteq F$ absolutely convex, $\sigma(F, E)$-compact. Then $B^{\circ\bullet} = b_2(B)$.*

Proof
It is easy to see that $B^\circ = b_2(B)^\bullet$. Also, the mapping $b_2 \colon F \to E^*$ is $\sigma(F, E)$-$\sigma(E^*, E)$-continuous, and this implies that $b_2(B)$ is $\sigma(E^*, E)$-compact, and therefore $\sigma(E^*, E)$-closed (and absolutely convex). Therefore the bipolar theorem in $\langle E, E^* \rangle$ yields $B^{\circ\bullet} = b_2(B)^{\bullet\bullet} = b_2(B)$. $\qquad\qquad\square$

Proof of Theorem 5.1
(a) As before, let

$$\mathcal{M}_\mu := \{ B \subseteq F ; \ B \text{ absolutely convex}, \ \sigma(F, E)\text{-compact}\}.$$

Then $\lambda B \in \mathcal{M}_\mu$ for all $B \in \mathcal{M}_\mu$, $\lambda \in \mathbb{K}$. Also, if $A, B \in \mathcal{M}_\mu$, then $C := A + B \in \mathcal{M}_\mu$, by Lemma 5.2, and $A \cup B \subseteq C$. These properties imply that

$$\mathcal{U} := \{ B^\circ ; \ B \in \mathcal{M}_\mu \}$$

is a neighbourhood base of zero for $\mu(E, F) = \tau_{\mathcal{M}_\mu}$, by Proposition 3.9(b). Applying Theorem 3.2 and Lemma 5.3, we conclude that

$$(E, \mu(E, F))' = \bigcup_{U \in \mathcal{U}} U^\bullet = \bigcup_{B \in \mathcal{M}_\mu} B^{\circ\bullet} = \bigcup_{B \in \mathcal{M}_\mu} b_2(B) = b_2(F),$$

where in the last equality we have used that $\text{aco}\{y\} \in \mathcal{M}_\mu$ for all $y \in F$, and therefore $\bigcup \mathcal{M}_\mu = F$.

(b) The necessity was shown at the end of Chapter 4. The sufficiency follows from Theorem 1.8 and part (a). $\qquad\qquad\square$

Remark 5.4 In the context of Theorem 5.1(b) one can ask whether there may exist non-locally convex linear topologies on E such that the dual is F. This will be answered affirmatively by the following examples. $\qquad\qquad\triangle$

Examples 5.5
(a) We present a dual pair $\langle E, F \rangle$ and a non-locally convex linear topology on E which is finer than the Mackey topology $\mu(E, F)$, but such that the dual space is still F.

Let $0 < p < 1$, $\ell_p := \{ x = (x_k) \in \mathbb{K}^{\mathbb{N}} ; \ \sum_{k=1}^\infty |x_k|^p < \infty \}$, and define the metric $d_p(x, y) := \sum_{k=1}^\infty |x_k - y_k|^p$ on ℓ_p. Then it can be shown that ℓ_p, with the topology defined by the metric d_p, is a topological vector space. Also, it is not difficult to show that $\ell_p \subseteq \ell_1$, with continuous inclusion. In contrast to the situation in Example 2.4, there are continuous linear functionals on ℓ_p, in fact $\ell_\infty = \ell_1' \subseteq \ell_p'$.

We will show that these are all continuous linear functionals, i.e., $\ell_p' = \ell_1'$. Let $\eta \in \ell_p'$. Then the restriction of η to c_c is of the form $\eta(x) = \sum_{k \in \mathbb{N}} \eta_k x_k$ ($x \in c_c$). The continuity of η implies that there exists $\delta > 0$ such that $|\eta(x)| \leqslant 1$ for all $x \in \ell_p$ with $\sum_{k \in \mathbb{N}} |x_k|^p \leqslant \delta$.

Applying this to multiples of unit vectors, we conclude that $1 \geqslant |\eta(\delta^{1/p} e_k)| = \delta^{1/p} |\eta_k|$ $(k \in \mathbb{N})$, and therefore $(\eta_k) \in \ell_\infty$. This shows that each continuous linear functional on ℓ_p is also continuous on ℓ_1.

Let $E := \ell_p$, let τ_p be the topology defined by the metric d_p, and let τ_1 be the topology defined by the norm $\|\cdot\|_1$. Then, because ℓ_p is dense in ℓ_1, we have shown that $(E, \tau_p)' = (E, \tau_1)'$. We will see in Chapter 6 that $\mu(E, E') = \tau_1$ (τ_1 being a locally convex metric topology). As the topology τ_p is strictly finer than τ_1 (the sequence $\left(n^{-1/p} \sum_{j=1}^{n} e_j\right)$ is a null sequence for τ_1, but not for τ_p), it follows from the Mackey–Arens theorem that τ_p is not a locally convex topology.

A more detailed investigation of linear topologies finer than the Mackey topology, but still resulting in the same dual space, has been given in [Kąk87].

(b) The second example on the issue of 'compatible' non-locally convex topologies on E, for a separating dual pair $\langle E, F \rangle$, concerns the question whether there also can be a non-locally convex topology between the weak topology and the Mackey topology. The author could not trace a treatment of this question in the literature. The example will show that the answer to this question is also positive.

Let $0 < q < 1 \leqslant p < \infty$, let $E := L_p(0, 1)$, and let σ be the weak topology. Define the metric $d_q(f, g) := \int |f - g|^q$ on E, and denote by τ_q the topology defined by this metric. (In other words, τ_q is the topology on E induced by the non-locally convex linear topology of $L_q(0, 1)$.) Define τ on E as the initial topology with respect to the mappings id: $E \to (E, \sigma)$ and id: $E \to (E, \tau_q)$. Then $\tau \supseteq \sigma$. But also $\mu(E, E') = \tau_{\|\cdot\|_p} \supseteq \tau$, because the mappings id: $(E, \|\cdot\|_p) \to (E, \sigma)$ and id: $(E, \|\cdot\|_p) \to (E, \tau_q)$ are continuous. (Again, for the fact that $\mu(E, E') = \tau_{\|\cdot\|_p}$ we refer to Chapter 6.)

It remains to verify that the topology τ is not locally convex. The first observation is that a neighbourhood base of zero for τ is given by $\{U_{F, \varepsilon}; \ F \subseteq (E, \|\cdot\|_p)' \text{ finite}, \ \varepsilon > 0\}$, where

$$U_{F, \varepsilon} := \{f \in E; \ \sup_{\eta \in F} |\eta(f)| \leqslant \varepsilon, \ d_q(f, 0) \leqslant \varepsilon\}.$$

We show that the τ-neighbourhood $U := \{f \in E; \ d_q(f, 0) < 1\}$ of zero does not contain the convex hull of any of the zero neighbourhoods $U_{F, \varepsilon}$. More precisely, we show that the convex hull of $U_{F, \varepsilon}$ contains elements $f \in \bigcap_{\eta \in F} \eta^{-1}(0)$ with $d_q(f, 0)$ arbitrarily large.

Let $F \subseteq (E, \|\cdot\|_p)'$ be finite, $\varepsilon > 0$. Let $n \in \mathbb{N}$. For $1 \leqslant j \leqslant n$ we can find $f_j \in \bigcap_{\eta \in F} \eta^{-1}(0) \cap L_p(\frac{j-1}{n}, \frac{j}{n})$ (where we identify $L_p(\frac{j-1}{n}, \frac{j}{n})$ with the set of elements in $L_p(0, 1)$ vanishing on $(0, 1) \setminus (\frac{j-1}{n}, \frac{j}{n})$), $\int |f_j|^q = \varepsilon$. Then $f := \frac{1}{n}(f_1 + \cdots + f_n) \in \text{co } U_{F, \varepsilon}$, and $d_q(f, 0) = \int |f|^q = n^{-p} \left(\int |f_1|^q + \cdots + \int |f_n|^q \right) = n^{1-q} \varepsilon$.

What we have shown is that U does not contain a convex neighbourhood of zero, and therefore the topology τ is not locally convex. $\triangle$

Example 5.6

As an illustration of the Mackey–Arens theorem we will compute the Mackey topology $\mu(\ell_\infty, \ell_1)$.

(i) First we determine the $\sigma(\ell_1, \ell_\infty)$-compact sets, i.e., the weakly compact sets of ℓ_1.

We suppose that it is known that a set $A \subseteq \ell_1$ is relatively compact if and only if A is bounded and $\sup_{x \in A} \sum_{j=n}^{\infty} |x_j| \to 0$ $(n \to \infty)$. (This is easy to show.) Equivalently, A is relatively compact if and only if there exists a decreasing null sequence $\alpha = (\alpha_n)$ in $(0, \infty)$ such that

$$A \subseteq A'_\alpha := \Big\{ x = (x_j)_j \in \ell_1 ;\ \sum_{j=n}^\infty |x_j| \leqslant \alpha_n\ (n \in \mathbb{N}) \Big\}.$$

In what follows now we will use the fact that A is weakly compact if and only if A is compact. This will be shown below in Corollary 5.10.

(ii) With $\mathcal{A} := \{ \alpha ;\ \alpha = (\alpha_j)_j$ a decreasing null sequence in $(0, \infty) \}$, $\mathcal{M}' := \{ A'_\alpha ;\ \alpha \in \mathcal{A} \}$ we now obtain

$$\tau_{\mathcal{M}'} = \mu(\ell_\infty, \ell_1)$$

(note that the sets A'_α are absolutely convex). Next we show that one can replace the collection $\mathcal{M}'$ by

$$\mathcal{M} := \{ \alpha B_{\ell_1} ;\ \alpha \in \mathcal{A} \},$$

where $\alpha B_{\ell_1} := \{ \alpha x := (a_j x_j)_j ;\ x \in \ell_1,\ \|x\| \leqslant 1 \}$.

For $\alpha \in \mathcal{A}$ one has $\alpha B_{\ell_1} \subseteq A'_\alpha$, because from $x \in B_{\ell_1}$ one obtains $\sum_{j=n}^\infty \alpha_j |x_j| \leqslant \alpha_n \sum_{j=n}^\infty |x_j| \leqslant \alpha_n\ (n \in \mathbb{N})$.

For $\alpha \in \mathcal{A}$ one has $\sum_{n=1}^\infty (\alpha_n - \alpha_{n+1}) = \alpha_1\ (< \infty)$. Therefore there exists $\beta \in \mathcal{A}$ such that $\sum_{n=1}^\infty \frac{1}{\beta_n} (\alpha_n - \alpha_{n+1}) \leqslant 1$ (see Remark 5.7 below). With this β one has $A'_\alpha \subseteq \beta B_{\ell_1}$, because for $x \in A'_\alpha$ one has

$$\sum_{n=1}^\infty \frac{|x_n|}{\beta_n} = \frac{1}{\beta_1} \sum_{n=1}^\infty |x_n| + \Big(\frac{1}{\beta_2} - \frac{1}{\beta_1} \Big) \sum_{n=2}^\infty |x_n| + \cdots$$

$$\leqslant \frac{1}{\beta_1} (\alpha_1 - \alpha_2) + \frac{1}{\beta_2} (\alpha_2 - \alpha_3) + \cdots \leqslant 1.$$

(iii) For $\alpha \in \mathcal{A}$ we set $A_\alpha := \alpha B_{\ell_1}$, $p_\alpha := q_{A_\alpha} (= p_{A_\alpha^\circ})$, and we compute p_α. For $y \in \ell_\infty$ one has

$$p_\alpha(y) = \sup_{x \in A_\alpha} |\langle y, x \rangle| = \sup_{x \in B_{\ell_1}} |\langle y, \alpha x \rangle| = \sup_{n \in \mathbb{N}} \alpha_n |y_n|.$$

With $P := \{ p_\alpha ;\ \alpha \in \mathcal{A} \}$ one now obtains $\mu(\ell_\infty, \ell_1) = \tau_P$.

Note that in the set P of norms there does not exist a sequence such that every norm in P is dominated by some norm in the sequence. This means that the topology $\mu(\ell_\infty, \ell_1)$ is not metrisable.

(iv) Finally we show that $(\ell_\infty, \tau_P)' = \ell_1$. (Well, we know from the Mackey–Arens theorem that this holds, but for the present example we want to obtain it directly.)

Let $\eta \in (\ell_\infty, \tau_P)'$. Since the set $\mathcal{A}$ is directed, there exists $\alpha \in \mathcal{A}$ such that

$$|\eta(y)| \leqslant p_\alpha(y) \quad (y \in \ell_\infty).$$

This means that $\eta \in (\ell_\infty, p_\alpha)'$. The mapping

$$j \colon (\ell_\infty, p_\alpha) \to c_0, \quad y \mapsto \alpha y$$

is isometric with dense range. Therefore there exists $\hat{\eta} \in \ell_1 = c_0'$ with

$$\eta(y) = \langle \alpha y, \hat{\eta} \rangle = \sum_{n=1}^\infty \alpha_n y_n \hat{\eta}_n,$$

i.e., η is represented by $x = (x_n)_n := (\alpha_n \hat{\eta}_n)_n \in \ell_1$ in the standard dual pairing of ℓ_1 and ℓ_∞. This shows that $(\ell_\infty, \tau_P)' \subseteq \ell_1$.

Also, for $x \in \ell_1$ there exists $\alpha \in \mathcal{A}$ such that $\left(\frac{1}{a_n} x_n\right)_n \in \ell_1$, by Remark 5.7. Therefore

$$\left| \sum_n y_n x_n \right| \leqslant \sum_n |y_n x_n| = \sum_n \left| \frac{x_n}{a_n} \right| \alpha_n |y_n| \leqslant \left\| \left(\frac{x_n}{a_n}\right) \right\|_1 p_\alpha(y) \quad (y \in \ell_\infty).$$

This shows that $(\ell_\infty, \tau_P)' = \ell_1$. $\quad\triangle$

Remark 5.7 Let $(a_n)_n$ be a sequence in $[0, \infty)$, $\sum_n a_n < \infty$. We show that there exists an increasing sequence $(c_n)_n$ in $(0, \infty)$, $c_n \to \infty$ $(n \to \infty)$, and such that $\sum_n c_n a_n < \infty$.

There exists a strictly increasing sequence $(n_j)_j$ in $\mathbb{N}$ such that

$$\sum_{n \geqslant n_j} a_n \leqslant \frac{1}{4^j} \quad (j \in \mathbb{N}).$$

Define

$$c_n := \begin{cases} 1 & \text{for } n < n_1, \\ 2^j & \text{for } n_j \leqslant n < n_{j+1}, \ j \in \mathbb{N}. \end{cases}$$

Then (c_n) is increasing, $c_n \to \infty$, and

$$\sum_n c_n a_n \leqslant \sum_{n=1}^{n_1-1} a_n + \sum_{j=1}^\infty 2^j \sum_{n=n_j}^\infty a_n \leqslant \sum_{n=1}^{n_1-1} a_n + \sum_{j=1}^\infty 2^j 4^{-j} < \infty. \quad\triangle$$

In the treatment of Example 5.6 we have used that compactness and weak compactness are equivalent for subsets of ℓ_1. This fact and related properties will be treated now.

Theorem 5.8

Let (x^n) be a $\sigma(\ell_1, \ell_\infty)$-Cauchy sequence in ℓ_1. Then (x^n) is convergent in norm.

By a $\sigma(\ell_1, \ell_\infty)$-**Cauchy sequence** or a weak Cauchy sequence $(x^n)_{n \in \mathbb{N}}$ in ℓ_1 we understand a sequence with the property that $(\langle x^n, y \rangle)_n$ is a Cauchy sequence in $\mathbb{K}$ for each $y \in \ell_\infty$; analogously for '$\sigma(\ell_1, c_0)$-Cauchy sequence'. This provisional definition is consistent with the definition of Cauchy sequences in topological vector spaces given later; see Chapter 9.

Proof of Theorem 5.8

The sequence (x^n) is bounded, by the uniform boundedness theorem (Theorem B.3). Therefore (x^n) has a $\sigma(\ell_1, c_0)$-cluster point $x \in \ell_1$, by Theorem 4.7, and because (x^n) is also a $\sigma(\ell_1, c_0)$-Cauchy sequence we conclude that $x^n \to x$ with respect to $\sigma(\ell_1, c_0)$. (To make this clear, let $F \subseteq c_0$ be a finite subset, and let $\varepsilon > 0$. Then there exists $n_0 \in \mathbb{N}$ such that $\sup_{y \in F} |\langle x^n - x^m, y \rangle| < \varepsilon$ $(m, n \geq n_0)$, and there exists $n_1 \geq n_0$ such that $\sup_{y \in F} |\langle x^{n_1} - x, y \rangle| < \varepsilon$. Then $\sup_{y \in F} |\langle x^n - x, y \rangle| < 2\varepsilon$ $(n \geq n_0)$. See also Remark 9.1(b).) Without loss of generality we may assume that $x = 0$.

We show that $x^n \to 0$ in ℓ_1 as $n \to \infty$. Assuming the contrary, we obtain $\varepsilon := \limsup_{n \to \infty} \|x^n\| > 0$.

Then it is not difficult to see that there exist a subsequence $(x^{n_k})_k$ and a subsequence $(m_k)_k$ of $\mathbb{N}$, such that

$$\sum_{j=1}^{m_k} |x_j^{n_k}| \leq \frac{\varepsilon}{8}, \quad \sum_{j=m_k+1}^{m_{k+1}} |x_j^{n_k}| \geq \frac{\varepsilon}{2}, \quad \sum_{j=m_{k+1}+1}^{\infty} |x_j^{n_k}| \leq \frac{\varepsilon}{8}.$$

For simplicity of notation and without loss of generality, we assume that $(x^{n_k}) = (x^k)$. We define $y = (y_j) \in \ell_\infty$ by

$$y_j := \begin{cases} 0 & \text{if } 1 \leq j \leq m_1, \\ (-1)^k \overline{\operatorname{sgn} x_j^k} & \text{if } m_k < j \leq m_{k+1}, \ k \in \mathbb{N}, \end{cases}$$

where the **signum function** $\operatorname{sgn} \colon \mathbb{K} \to \mathbb{K}$ is defined by $\operatorname{sgn} \lambda := \lambda/|\lambda|$ if $\lambda \neq 0$, $\operatorname{sgn} 0 := 0$. For $k \geq 2$ one then obtains

$$\langle x^k - x^{k-1}, y \rangle$$

$$= \sum_1^{m_k} x_j^k y_j + (-1)^k \sum_{m_k+1}^{m_{k+1}} |x_j^k| + \sum_{m_{k+1}+1}^{\infty} x_j^k y_j$$

$$- \sum_1^{m_{k-1}} x_j^{k-1} y_j - (-1)^{k-1} \sum_{m_{k-1}+1}^{m_k} |x_j^{k-1}| - \sum_{m_k+1}^{\infty} x_j^{k-1} y_j,$$

and this implies that

$$|\langle x^k - x^{k-1}, y \rangle| = |(-1)^k \langle x^k - x^{k-1}, y \rangle| \geqslant -\frac{\varepsilon}{8} + \frac{\varepsilon}{2} - \frac{\varepsilon}{8} - \frac{\varepsilon}{8} + \frac{\varepsilon}{2} - \frac{\varepsilon}{8} = \frac{\varepsilon}{2}.$$

This is a contradiction to (x^k) being a weak Cauchy sequence. □

Remarks 5.9 (a) Of course, the weak Cauchy sequence in Theorem 5.8 also converges weakly to $\lim x^n$. So, Theorem 5.8 implies that ℓ_1 is weakly sequentially complete (in the terminology of Chapter 9). In Chapter 15, this property will be extended to all L_1-spaces.

(b) The method employed in the proof of Theorem 5.8 is known under the name of 'sliding hump' or 'gliding hump' method. △

Corollary 5.10 *A set $A \subseteq \ell_1$ is weakly compact if and only if A is compact.*

Proof
It is clear that A is weakly compact if A is compact. Now assume that A is weakly compact, and let (x^n) be a sequence in A. Proposition 5.11, proved below, then implies that (x^n) contains a weakly convergent subsequence; for the application of Proposition 5.11 note that ℓ_1 is separable. Using Theorem 5.8 we deduce that this subsequence is also convergent in ℓ_1. So it is shown that A is sequentially compact, and because A is a metric space, this implies that A is compact. □

Proposition 5.11 *Let E be a separable Banach space, and let $A \subseteq E$ be weakly compact. Then $(A, \sigma(E, E') \cap A)$ is metrisable.*

Proof
First we show that there exists a sequence $(x_n')_n$ in E' separating the points of E. Let $(x_n)_n$ be a dense sequence in the unit sphere S_E of E. The Hahn–Banach theorem implies that there exists a sequence (x_n') in $S_{E'}$ such that $x_n'(x_n) = 1$ $(n \in \mathbb{N})$. Then the set $\{x_n' ; \, n \in \mathbb{N}\}$ separates the points of E, because for all $x \in S_E$ there exists $n \in \mathbb{N}$ such that $\|x - x_n\| < 1$, and therefore $x_n'(x) = x_n'(x - x_n) + x_n'(x_n) \neq 0$.

Let ρ be the initial topology on E with respect to the sequence $\big(E \ni x \mapsto \langle x, x_n' \rangle \in \mathbb{K}\big)_{n \in \mathbb{N}}$. Then Proposition 2.17 implies that ρ is metrisable. On A one obtains $\rho \cap A = \sigma(E, E') \cap A$, because ρ is Hausdorff and coarser than $\sigma(E, E')$; recall Lemma 4.12. □

Notes Theorem 5.1 was proved by Mackey and Arens. More precisely, Mackey [Mac46, Theorem 5] showed that among the compatible locally convex topologies there is a coarsest one (the weak topology) and a finest one; and Arens [Are47, Theorem 2] provided the description of the finest topology as what is now called 'Mackey topology'. In Example 5.6 we present an example for the Mackey topology (in a non-reflexive situation), where one can still compute everything. It turns out that, in order to carry this out for the dual pair $\langle \ell_\infty, \ell_1 \rangle$, one needs interesting properties of convergence in ℓ_1, which are singled out in Theorem 5.8 and Corollary 5.10; these properties will also be needed later in Chapter 15.

Topologies on E'', Quasi-barrelled and Barrelled Spaces

The topics of this chapter draw their motivation, with a locally convex space E, from two questions: find topologies on E'' such that the canonical mapping $\kappa\colon E \to E''$ is continuous, and investigate properties of topologies on E ensuring that κ is continuous, if E'' is provided with the strong topology $\beta(E'', E')$. The first issue leads to the 'natural topology' on E'', the second leads to 'quasi-barrelled' spaces, and in particular, the answer to the second question motivates the investigation of further related properties of locally convex spaces.

The following theorem on bounded sets is important for the subsequent treatment.

Theorem 6.1 (Mackey)
Let E be a locally convex space, $A \subseteq E$. Then A is bounded if and only if A is weakly bounded.

Note that the theorem could have been formulated equivalently for a dual pair $\langle E, F \rangle$, by stating that a set $A \subseteq E$ is bounded for some compatible locally convex topology if and only if A is $\sigma(E, F)$-bounded.

The next two lemmas are preparations for the proof. The first of these is the crucial reduction of the theorem to the uniform boundedness theorem.

Lemma 6.2 *Let (E, p) be a semi-normed space, $A \subseteq E$. Then A is bounded if and only if A is weakly bounded.*

Proof
The necessity follow from the continuity of id$\colon (E, p) \to (E, \sigma(E, E'))$ and Lemma 3.4(a). To show the sufficiency, we note that $(E, p)'$ is a Banach space, with norm

$$\|x'\| = \sup\{|\langle x, x'\rangle|\,;\ x \in E,\ p(x) \leqslant 1\}.$$

© Springer Nature Switzerland AG 2020
J. Voigt, *A Course on Topological Vector Spaces*, Compact Textbooks in Mathematics,
https://doi.org/10.1007/978-3-030-32945-7_6

For the canonical map $\kappa \colon E \to E''$ one obtains $\|\kappa(x)\|_{E''} = p(x)$ $(x \in E)$ as a consequence of the Hahn–Banach theorem; see Corollary A.4. The weak boundedness of A implies that $\kappa(A)$ is $\sigma(E'', E')$-bounded; hence the uniform boundedness theorem, Theorem B.3, implies that $\kappa(A)$ is bounded. Therefore $\sup_{x \in A} p(x) < \infty$, i.e., A is bounded. $\qquad\square$

Lemma 6.3 *Let E, F be locally convex spaces, $f \colon E \to F$ linear and continuous. Then f is $\sigma(E, E')$-$\sigma(F, F')$-continuous.*

Proof
Theorem 1.2 implies that it is sufficient to show that $y' \circ f$ is $\sigma(E, E')$-continuous for all $y' \in F'$. This, however, is true because $y' \circ f \in E'$. $\qquad\square$

Proof of Theorem 6.1
The necessity follows from Lemma 3.4(a). For the sufficiency recall that E carries the initial topology with respect to $(\mathrm{id} \colon E \to (E, p))_{p \in P}$, where P is a set of semi-norms (Corollary 2.15). For $p \in P$ the identity $\mathrm{id} \colon E \to (E, p)$ is continuous, so that Lemma 6.3 implies that A is weakly bounded in (E, p), and therefore bounded in (E, p), by Lemma 6.2. Therefore Lemma 3.4(b) implies that A is bounded. $\qquad\square$

We recall that, for a locally convex space E, the bidual is defined as $E'' = (E', \beta(E', E))'$. The following considerations are motivated by the question for a topology on E'' inducing the original topology on E under the canonical map $\kappa \colon E \to E''$; the answer will be given in Theorem 6.7. A related question is finding properties of the topology of E such that the canonical map is continuous if E'' is provided with the strong topology $\beta(E'', E')$. The answer will given in Theorem 6.8: E should be 'quasi-barrelled'. This investigation opens up the discussion of further interesting notions important for locally convex spaces.

Proposition 6.4 *Let E be a locally convex space, $\mathcal{U}$ the collection of closed absolutely convex neighbourhoods of zero, and define*

$$\mathcal{M}_{\mathrm{n}} := \{U^\circ; \; U \in \mathcal{U}\}.$$

Then the topology of E is the polar topology $\tau_{\mathcal{M}_{\mathrm{n}}}$ (in the dual pair $\langle E, E' \rangle$).

Proof
Recall from Theorem 2.14 that $\mathcal{U}$ is a neighbourhood base of zero. Then Theorem 3.6 and Proposition 3.9(b) imply that

$$\mathcal{U} = \{U^{\circ\circ}; \; U \in \mathcal{U}\} = \{B^\circ; \; B \in \mathcal{M}_{\mathrm{n}}\}$$

is also a neighbourhood base of zero for $\tau_{\mathcal{M}_{\mathrm{n}}}$. $\qquad\square$

Remarks 6.5 (a) With the collection $\mathcal{M}_n$ we want to define a polar topology on E''. To verify that this is possible one has to show that $\mathcal{M}_n \subseteq \mathcal{B}_\sigma(E', E'')$; this will be done subsequently in Proposition 6.6.

(b) Let E be a topological vector space, $B \subseteq E'$. Then B is equicontinuous at 0 if for all $\varepsilon > 0$ there exists $U \in \mathcal{U}_0(E)$ such that $|\langle x, x'\rangle| \leqslant \varepsilon$ ($x \in U$, $x' \in B$), or equivalently, if there exists $U \in \mathcal{U}_0(E)$ such that $|\langle x, x'\rangle| \leqslant 1$ ($x \in U$, $x' \in B$), i.e., such that $B \subseteq U^\circ$. For simplicity, we will then call B **equicontinuous**.

Because of linearity, a set $B \subseteq E'$ is equicontinuous at 0 if and only if it is 'uniformly equicontinuous', i.e., if for all $\varepsilon > 0$ there exists $U \in \mathcal{U}_0(E)$ such that $|\langle x, x'\rangle - \langle y, x'\rangle| = |\langle x - y, x'\rangle| \leqslant \varepsilon$ for all $x, y \in E$ with $x - y \in U$ and all $x' \in B$.

(c) Now let E be a locally convex space. In view of the preceding discussion, in Proposition 6.4 the collection $\mathcal{M}_n$ could have been replaced by the collection of all equicontinuous subsets of E' (called $\mathcal{E}$ in a discussion at the end of the present chapter). △

Let E be a vector space, $A, B \subseteq E$. We say that A **absorbs** B, or that B **is absorbed by** A, if there exists $\alpha > 0$ such that $B \subseteq \lambda A$ for all $\lambda \in \mathbb{K}$ with $|\lambda| \geqslant \alpha$.

Proposition 6.6 *Let E be a locally convex space, $U \in \mathcal{U}_0(E)$. Then U° is $\beta(E', E)$-bounded.*

Proof

Let $V \subseteq E'$ be a $\beta(E', E)$-neighbourhood of zero; without loss of generality we can assume that $V = B^\circ$ for some weakly bounded set $B \subseteq E$. Theorem 6.1 implies that B is bounded, and therefore U absorbs B. This, in turn, implies that $V = B^\circ$ absorbs U°. □

Let E be a locally convex space. With $\mathcal{M}_n$ from Proposition 6.4 the topology

$$\tau_n := \tau_{\mathcal{M}_n}$$

on E'', with the polar topology formed in the dual pair $\langle E'', E'\rangle$, is called the **natural topology**; this name is the motivation for the index 'n' in $\mathcal{M}_n$ and τ_n. (With the terminology 'natural topology' we follow [Köt66, § 23.4], [Sch71, Chap. IV, § 5.3].)

Proposition 6.6 can also be expressed by $\mathcal{M}_n \subseteq \mathcal{B}_\sigma(E', E'')$, and this implies that $\tau_n \subseteq \beta(E'', E')$.

Theorem 6.7

Let E be a locally convex space. Then E carries the initial topology with respect to the canonical map $\kappa \colon E \to (E'', \tau_n)$. The topology τ_n is the finest polar topology on E'' in the dual pair $\langle E'', E'\rangle$ for which $\kappa \colon E \to E''$ is continuous. If E is Hausdorff, then κ is an isomorphism from E to the subspace $\kappa(E)$ of (E'', τ_n).

Proof

In this proof the polars taken in $\langle E'', E' \rangle$ will be denoted by '$\bullet$' (whereas '$\circ$' denotes polars in $\langle E, E' \rangle$).

Let $U \in \mathcal{U}_0(E)$ be absolutely convex and closed. Then one has $\kappa^{-1}(U^{\circ\bullet}) = U^{\circ\circ} = U$. This shows that $\kappa \colon E \to (E'', \tau_n)$ is continuous and that E carries the initial topology as asserted. If E is Hausdorff, then κ is injective and one obtains $\kappa(U) = \kappa(\kappa^{-1}(U^{\circ\bullet})) = U^{\circ\bullet} \cap \kappa(E)$, and this implies the last assertion of the theorem.

Let $\tau_\mathcal{M}$ be a polar topology on E'', with $\mathcal{M} \subseteq \mathcal{B}_\sigma(E', E'')$, and assume that $\kappa \colon E \to (E'', \tau_\mathcal{M})$ is continuous. Let $B \in \mathcal{M}$. Then $B^\bullet$ is a $\tau_\mathcal{M}$-neighbourhood of zero, and by hypothesis $B^\circ = \kappa^{-1}(B^\bullet)$ is a neighbourhood of zero in E. Therefore $B \subseteq B^{\circ\circ} \in \mathcal{M}_n$, and this implies that $\tau_\mathcal{M} \subseteq \tau_n$. □

Theorem 6.8

Let E be a locally convex space. Then the following properties are equivalent:

  (i) $\tau_n = \beta(E'', E')$;

  (ii) *every $\beta(E', E)$-bounded set $B \subseteq E'$ is equicontinuous;*

  (iii) *if $U \subseteq E$ is absolutely convex, closed and **bornivorous** (i.e., U absorbs all bounded sets), then U is a neighbourhood of zero.*

Proof

As above, the polars taken in $\langle E'', E' \rangle$ will be denoted by '$\bullet$'.

(i) $\Rightarrow$ (ii). Let $B \subseteq E'$ be $\beta(E', E)$-bounded. Then B is $\sigma(E', E'')$-bounded, because $\beta(E', E) \supseteq \sigma(E', E'')$. Hence $B^\bullet$ is a $\beta(E'', E')$-neighbourhood of zero, and therefore $B^\circ = \kappa^{-1}(B^\bullet)$ is a neighbourhood of zero in E, i.e., B is equicontinuous.

(ii) $\Rightarrow$ (iii). Let U be as in (iii). It is sufficient to show that U° is equicontinuous. (Then $U = U^{\circ\circ}$ is a neighbourhood of zero.)

Let $V \subseteq E'$ be a $\beta(E', E)$-neighbourhood of zero, without restriction $V = B^\circ$, with bounded $B \subseteq E$. Then U absorbs B, and therefore $V = B^\circ$ absorbs U°. This shows that U° is $\beta(E', E)$-bounded, and property (ii) implies that U° is equicontinuous.

(iii) $\Rightarrow$ (i). By Theorem 6.7 it is sufficient to show that the canonical map $\kappa \colon E \to (E'', \beta(E'', E'))$ is continuous.

Let $U \subseteq E''$ be a $\beta(E'', E')$-neighbourhood of zero, without restriction $U = B^\bullet$ with a $\sigma(E', E'')$-bounded set $B \subseteq E'$. We have to show that $\kappa^{-1}(U) = \kappa^{-1}(B^\bullet) = B^\circ$ is a neighbourhood of zero. Because of (iii) it is sufficient to show that B° is bornivorous.

Let $A \subseteq E$ be bounded. Then A° is a $\beta(E', E)$-neighbourhood of zero, therefore A° absorbs B (where it was used that Theorem 6.1 implies that B is $\beta(E', E)$-bounded). This implies that $A \subseteq A^{\circ\circ}$ is absorbed by B°. □

Let E be a locally convex space.

A set $B \subseteq E$ is a **barrel** if B is absolutely convex, closed and absorbing. Obviously, B is a barrel if and only if there exists a $\sigma(E', E)$-bounded set $A \subseteq E'$ such that $B = A^\circ$. The space E is **barrelled** if every barrel is a neighbourhood of zero. E is

quasi-barrelled (also 'infrabarrelled') if every bornivorous barrel is a neighbourhood of zero. We note that Theorem 6.8 characterises quasi-barrelled spaces.

Theorem 6.9

Let E be a locally convex space and a Baire space (see Appendix B). Then E is barrelled. In particular, Fréchet spaces and Banach spaces are barrelled.

Proof

Let $B \subseteq E$ be a barrel. Then $E = \bigcup_{n \in \mathbb{N}} nB$, and therefore $\overset{\circ}{B} \neq \varnothing$. Let $x \in \overset{\circ}{B}$, U an absolutely convex neighbourhood of zero such that $x + U \subseteq B$. Then $-x + U = -(x + U) \subseteq B$ as well, and

$$U = \tfrac{1}{2}U + \tfrac{1}{2}U = \tfrac{1}{2}(x + U) + \tfrac{1}{2}(-x + U) \subseteq B. \qquad \square$$

Examples 6.10

(a) The spaces $C(X)$, for X σ-compact Hausdorff locally compact, and $C^k(\Omega)$, for $\Omega \subseteq \mathbb{R}^n$ open, $k \in \mathbb{N}_0 \cup \{\infty\}$ are barrelled. Also $L_{p,\mathrm{loc}}(\Omega)$, for $\Omega \subseteq \mathbb{R}^n$ open, $1 \leqslant p \leqslant \infty$, is a barrelled space.

(b) The space $(c_c, \|\cdot\|_\infty)$ is an example of a quasi-barrelled space that is not barrelled. Indeed, let $(\alpha_n)_n$ be a null sequence in $(0, \infty)$. Then

$$B := \big\{(x_n)_n \,;\, |x_n| \leqslant \alpha_n \ (n \in \mathbb{N})\big\}$$

is a barrel, but not a neighbourhood of zero. Thus the space is not barrelled. However, the unit ball is bounded, therefore every bornivorous set absorbs the unit ball and therefore is a neighbourhood of zero; hence the space is quasi-barrelled. $\qquad \triangle$

A topological vector space is called **bornological** if it is locally convex, and every absolutely convex bornivorous set is a neighbourhood of zero. Evidently, every bornological space is quasi-barrelled.

Proposition 6.11 *Let E be a metrisable locally convex space. Then E is bornological.*

Proof

There exists a decreasing neighbourhood base of zero $(U_n)_n$ in E. Let $A \subseteq E$ be a subset which is not a neighbourhood of zero. Then for all $n \in \mathbb{N}$ there exists $x_n \in U_n \setminus nA$. Hence, (x_n) is a null sequence, therefore bounded, but the set $\{x_n \,;\, n \in \mathbb{N}\}$ is not absorbed by A. Thus, A is not bornivorous.

This shows that *every* bornivorous subset of E is a neighbourhood of zero. $\qquad \square$

The reader may have noticed that the author distinguishes carefully between sets and families. (A sequence of functions, for instance, which is a family, would always be written as $(f_n)_{n\in\mathbb{N}}$ or $(f_n; n \in \mathbb{N})$, maybe sometimes simply as (f_n), but never as $\{f_n\}_{n\in\mathbb{N}}$.) A neighbourhood base of zero is, by definition, a collection (i.e., a set) of sets. Nevertheless it is convenient to write a countable neighbourhood base of zero as a sequence $(U_n)_{n\in\mathbb{N}}$, as above, in particular if one wants to say that it is 'decreasing'. This kind of 'inconsistency' should not lead to confusion; the corresponding remark applies also to the notion 'cobase of bounded sets', defined in Chapter 7.

Proposition 6.12 *Let (E, τ) be a quasi-barrelled locally convex space. Then $\tau = \mu(E, E')$.*

Proof

'$\subseteq$' is obvious.

For '$\supseteq$' let U be a $\mu(E, E')$-neighbourhood of zero, without restriction a barrel. Then U is bornivorous because of Mackey's theorem (Theorem 6.1). Since E is quasi-barrelled, it follows that U is a neighbourhood of zero.   □

Remark 6.13 Locally convex spaces (E, τ) such that $\tau = \mu(E, E')$ are also called **Mackey spaces**. However, we warn the reader that in some references bornological spaces are called Mackey spaces.   △

Theorem 6.14

Let (E, τ) be a locally convex space. Then the following statements are equivalent.
  (i)   *E is barrelled;*
  (ii) *every $\sigma(E', E)$-bounded set $B \subseteq E'$ is equicontinuous;*
  (iii) *$\tau = \beta(E, E')$.*

Proof

(i) $\Rightarrow$ (ii). Let $B \subseteq E'$ be $\sigma(E', E)$-bounded. Then B° is a barrel, and therefore a neighbourhood of zero. The latter is equivalent to B being equicontinuous.

(ii) $\Rightarrow$ (iii). Since equicontinuous subsets of E' are $\sigma(E', E)$-bounded (a consequence of Lemma 3.4(b)), one could formulate (ii) also by saying that the collection of $\sigma(E', E)$-bounded sets is identical to the collection of equicontinuous sets. Proposition 6.4 implies that τ is the polar topology corresponding to the collection of equicontinuous subsets of E'; hence it follows that $\tau = \beta(E, E')$.

(iii) $\Rightarrow$ (i). If $B \subseteq E$ is a barrel, then B° is $\sigma(E', E)$-bounded, and therefore $B = (B^\circ)^\circ$ is a $\beta(E, E')$-neighbourhood of zero.   □

Remark 6.15 The statement in (ii) of Theorem 6.14 could be interpreted as a kind of uniform boundedness theorem (for linear functionals).   △

In order to illustrate the interplay between the different notions we have introduced we define the following notation. For a locally convex space E let

$$\mathcal{E} := \big\{ B \subseteq E'; \ B \text{ equicontinuous} \big\},$$

$$\mathcal{C} := \big\{ B \subseteq E'; \ \overline{\text{aco } B}^{\,\sigma(E',E)} \ \sigma(E', E)\text{-compact} \big\},$$

$$\mathcal{B}_\beta := \big\{ B \subseteq E'; \ B \ \beta(E', E)\text{-bounded} \big\},$$

$$\mathcal{B}_\sigma := \big\{ B \subseteq E'; \ B \ \sigma(E', E)\text{-bounded} \big\}.$$

Lemma 6.16 $\mathcal{E} \subseteq \mathcal{C} \subseteq \mathcal{B}_\beta \subseteq \mathcal{B}_\sigma.$

Proof

The first inclusion is the Alaoglu–Bourbaki theorem (Theorem 4.7). For the second inclusion note that $\tau_\mathcal{C} = \mu(E', E)$. Therefore, for a set $C = C^{\circ\circ} \in \mathcal{C}$ the polar C° is a $\mu(E, E')$-neighbourhood of zero, absorbing each bounded set $B \subseteq E$ (by Theorem 6.1), which implies that B° absorbs $C^{\circ\circ} = C$. As the sets B° constitute a $\beta(E', E)$-neighbourhood base of zero, it follows that C is $\beta(E', E)$-bounded. The third inclusion is clear because $\beta(E', E) \supseteq \sigma(E', E)$. $\qquad\square$

Remark 6.17 For a locally convex space (E, τ) we summarise the obtained results in a scheme of implications:

$$\mathcal{E} = \mathcal{C} = \mathcal{B}_\beta \quad \Longrightarrow \quad \mathcal{E} = \mathcal{C}$$

$$\Updownarrow \qquad\qquad\qquad \Updownarrow$$

$$\text{metrisable} \ \Longrightarrow \ \text{bornological} \ \Longrightarrow \ \text{quasi-barrelled} \ \Longrightarrow \ \tau = \mu(E, E')$$

$$\Uparrow \qquad\qquad\qquad\qquad \Uparrow$$

$$\text{Fréchet space} \ \Longrightarrow \ \text{Baire} \ \Longrightarrow \ \text{barrelled}$$

$$\Updownarrow$$

$$\tau = \beta(E, E')$$

$$\Updownarrow$$

$$\mathcal{E} = \mathcal{C} = \mathcal{B}_\beta = \mathcal{B}_\sigma \qquad\qquad \triangle$$

Closing the present chapter, we give a statement concerning bornological spaces. The equivalence expressed in condition (ii) gives a hint for the name 'bornological': The continuity of a linear mapping is determined by the behaviour of the mapping on bounded sets.

Proposition 6.18 *Let E be a locally convex space. Then the following statements are equivalent:*

(i) *E is bornological;*

(ii) *for each locally convex space (resp. semi-normed space) F every **bounded** linear mapping $f: E \to F$ is continuous (where 'bounded' means that for all bounded $A \subseteq E$ the image $f(A)$ is bounded).*

Proof

(i) $\Rightarrow$ (ii) (for 'locally convex'). Let F be a locally convex space, $f: E \to F$ linear and bounded. Let $V \subseteq F$ be an absolutely convex neighbourhood of zero, $A \subseteq E$ bounded. Then $f(A)$ is bounded, and therefore is absorbed by V. Therefore $A \subseteq f^{-1}(f(A))$ is absorbed by $f^{-1}(V)$. This shows that the (absolutely convex!) set $f^{-1}(V)$ is bornivorous, hence a neighbourhood of zero.

(ii) (with 'semi-normed') $\Rightarrow$ (i). Let $U \subseteq E$ be absolutely convex and bornivorous, p_U the Minkowski functional of U. Then id: $E \to (E, p_U)$ is bounded: If A is bounded, then A is absorbed by $U \subseteq \{x \in E;\ p_U(x) \leqslant 1\}$. The hypothesis implies that id: $E \to (E, p_U)$ is continuous, and therefore U is a neighbourhood of zero in E. □

Notes Theorem 6.1 is due to Mackey [Mac46, Theorem 7]. Discussions on the natural topology can be found in [Köt66, V, § 23.4], [Sch71, Chap. IV, § 5.3]. The remaining topics of the chapter are rather standard and difficult to trace to the origins.

Fréchet Spaces and DF-Spaces

Besides Hilbert spaces and Banach spaces occurring as function spaces in analysis, an important role is also played by Fréchet spaces. It is for this reason that we include a chapter on some properties of metrisable locally convex spaces and Fréchet spaces. The first part of the chapter concerns the duality of Fréchet spaces: in short and simplified, duals of Fréchet spaces are DF-spaces, and duals of DF-spaces are Fréchet spaces. Looking at examples of duals of Fréchet spaces, one realises that quite often they can only be described as quotients, and this is the reason for inserting a short interlude on final topologies and topologies on quotient spaces. The third topic is a peculiarity of Fréchet spaces: They could also have been defined as 'completely metrisable' locally convex spaces.

For brevity it will be convenient to introduce the following notions. A **countably quasi-barrelled** (also 'countably infrabarrelled') space is a locally convex space with the property that each bornivorous countable intersection of closed absolutely convex neighbourhoods of zero is a neighbourhood of zero. It is evident that 'quasi-barrelled' implies 'countably quasi-barrelled'.

A **cobase of bounded sets** in a topological vector space E is a collection $\mathcal{B}$ of bounded sets with the property that for each bounded set $A \subseteq E$ there exists $B \in \mathcal{B}$ such that $A \subseteq B$.

A **DF-space** is a countably quasi-barrelled locally convex space possessing a countable cobase of bounded sets. We will show that the dual of a metrisable locally convex space is a DF-space and that the dual of a DF-space is a Fréchet space. Just for an easy example: Each normed space is a DF-space; it is bornological, hence countably quasi-barrelled, and the sequence $(B(0, n))_{n \in \mathbb{N}}$ is a countable cobase of bounded sets.

It will be convenient to use the notation E'_β for the dual of a locally convex space E, provided with the topology $\beta(E', E)$.

© Springer Nature Switzerland AG 2020

J. Voigt, *A Course on Topological Vector Spaces*, Compact Textbooks in Mathematics, https://doi.org/10.1007/978-3-030-32945-7_7

Theorem 7.1

Let E be a metrisable locally convex space, and let $(U_n)_{n \in \mathbb{N}}$ be a neighbourhood base of zero. Then $(U_n^\circ)_{n \in \mathbb{N}}$ is a countable cobase of bounded sets in E'_β, and the space E'_β is a DF-space.

Proof

Let $A \subseteq E'$ be a $\beta(E', E)$-bounded set. Then for each bounded (equivalently, $\sigma(E, E')$-bounded) absolutely convex set $B \subseteq E$ there exists $\lambda > 0$ such that $A \subseteq \lambda B^\circ$, therefore $A^\circ \supseteq \frac{1}{\lambda} B$. This means that A° is bornivorous, hence a neighbourhood of zero (because E is bornological, see Proposition 6.11), $A^\circ \supseteq U_n$ for some $n \in \mathbb{N}$, and $A \subseteq (A^\circ)^\circ \subseteq U_n^\circ$. This proves the first assertion of the theorem.

For the proof that E'_β is countably quasi-barrelled, let $(V_n)_{n \in \mathbb{N}}$ be a sequence of closed absolutely convex neighbourhoods of zero in E'_β, and suppose that $V := \bigcap_n V_n$ is bornivorous. (The 'problem' – see below – is that the sets V_n are not necessarily $\sigma(E', E)$-closed.) The procedure of the proof is to construct a sequence $(W_n)_{n \in \mathbb{N}}$ of absolutely convex $\sigma(E', E)$-closed neighbourhoods of zero in E' with $W_n \subseteq V_n$ ($n \in \mathbb{N}$) and such that $W := \bigcap_n W_n$ is still absorbing (even bornivorous). Then $W = (W^\circ)^\circ$, and W° is $\sigma(E, E')$-bounded; hence $W \subseteq V$ will be a neighbourhood of zero, and the proof will be finished.

For the construction of the sequence (W_n) let $A_n := U_n^\circ$ ($n \in \mathbb{N}$). We show by induction that there exist sequences (λ_n) in $(0, \infty)$ and (W_n) of absolutely convex $\sigma(E', E)$-closed neighbourhoods of zero satisfying

$$\lambda_n A_n \subseteq (\tfrac{1}{2} V) \cap W_j \qquad (1 \leqslant j < n), \tag{7.1}$$

$$\lambda_j A_j \subseteq W_n \subseteq V_n \qquad (1 \leqslant j \leqslant n) \tag{7.2}$$

for all $n \in \mathbb{N}$. Assume that $\lambda_1, \ldots, \lambda_{n-1}, W_1, \ldots, W_{n-1}$ are found. As V absorbs A_n, and $\bigcap_{1 \leqslant j < n} W_j$ is a neighbourhood of zero, there exists $\lambda_n > 0$ such that (7.1) is satisfied. The set

$$C_n := \mathrm{aco} \Big(\bigcup_{1 \leqslant j \leqslant n} \lambda_j A_j \Big) = \mathrm{aco} \Big(\bigcup_{1 \leqslant j \leqslant n} \lambda_j U_j^\circ \Big)$$

is $\sigma(E', E)$-compact by the Alaoglu–Bourbaki theorem (Theorem 4.7) and Lemma 7.2 below. As V_n is a neighbourhood of zero in E'_β, there exists a bounded set $B_n \subseteq E$ such that $B_n^\circ \subseteq \frac{1}{2} V_n$. Setting $W_n := B_n^\circ + C_n$ we obtain an absolutely convex $\sigma(E', E)$-closed neighbourhood of zero in E' satisfying $W_n \subseteq \frac{1}{2} V_n + \frac{1}{2} V \subseteq V_n$. (For '$\sigma(E', E)$-closed' we refer to Lemma 7.3 below.)

From (7.1) and (7.2) it follows that $\lambda_j A_j \subseteq W_n$ for all $j, n \in \mathbb{N}$, hence $\lambda_j A_j \subseteq W$ for all $j \in \mathbb{N}$, and this implies that W is bornivorous. □

In the proof given above we have used two properties concerning compact sets we will show now.

Lemma 7.2 *Let E be a topological vector space, and let $A_1, \ldots, A_n \subseteq E$ be compact absolutely convex sets. Then the set $\mathrm{aco}(A_1 \cup \cdots \cup A_n)$ is compact.*

Proof

It is sufficient to show this for $n = 2$. In this case one has

$$\mathrm{aco}(A_1 \cup A_2) = \left\{ \lambda_1 a_1 + \lambda_2 a_2 ; \; \lambda_1, \lambda_2 \geqslant 0, \; \lambda_1 + \lambda_2 = 1, \; a_1 \in A_1, \; a_2 \in A_2 \right\}.$$

Indeed, it is obvious that the right-hand side is balanced, it is easy to show that it is convex, and it clearly is the smallest convex set containing $A_1 \cup A_2$. This means that $\mathrm{aco}(A_1 \cup A_2)$ is the image of the compact set

$$\left\{ (\lambda_1, \lambda_2) \in [0, 1]^2 ; \; \lambda_1 + \lambda_2 = 1 \right\} \times A_1 \times A_2$$

under the continuous mapping $(\lambda_1, \lambda_2, a_1, a_2) \mapsto \lambda_1 a_2 + \lambda_2 a_2$; hence it is a compact set. $\quad\square$

Lemma 7.3 *Let E be a topological vector space, $A, B \subseteq E$, A closed, B compact.*
(a) *If $A \cap B = \varnothing$, then there exists $U \in \mathcal{U}_0(E)$ such that $A \cap (B + U) = \varnothing$.*
(b) *The set $A + B$ is closed.*

Proof

(a) For each $b \in B$ there exists $V_b \in \mathcal{U}_0$ such that $A \cap (b + V_b) = \varnothing$, and there exists an open $U_b \in \mathcal{U}_0$ such that $U_b + U_b \subseteq V_b$. For the open covering $(b + U_b)_{b \in B}$ of B there exists a finite subcovering $(b + U_b)_{b \in F}$. With $U := \bigcap_{b \in F} U_b \in \mathcal{U}_0(E)$ we then obtain

$$A \cap (B + U) \subseteq A \cap \left(\bigcup_{b \in F} (b + U_b) + U \right) \subseteq A \cap \left(\bigcup_{b \in F} (b + V_b) \right) = \varnothing.$$

(b) Let $x \in E \setminus (A + B)$, i.e., $(x - A) \cap B = \varnothing$. As $x - A$ is closed, part (a) implies that there exists $U \in \mathcal{U}_0$ such that $(x - A) \cap (B + U) = \varnothing$, i.e., $(x - U) \cap (A + B) = \varnothing$. This shows that $E \setminus (A + B)$ is open. $\quad\square$

The following lemma is a preparation for the description of the dual of DF-spaces.

Lemma 7.4 *Let E be a countably quasi-barrelled locally convex space, and let $B \subseteq E'$ be a $\beta(E', E)$-bounded countable union of equicontinuous subsets of E'. Then B is equicontinuous.*

Proof

Recall that B is equicontinuous if and only if $B^\circ \in \mathcal{U}_0(E)$.

If $C \subseteq E$ is bounded, then C° is a neighbourhood of zero for $\beta(E', E)$; hence there exists $\lambda > 0$ such that $\lambda C^\circ \supseteq B$, which implies $\frac{1}{\lambda} C \subseteq \frac{1}{\lambda} C^{\circ\circ} \subseteq B^\circ$. This shows that B° is bornivorous. Let (B_n) be a sequence of equicontinuous subsets of E' such that $B = \bigcup_n B_n$. Then $B^\circ = \bigcap_n B_n^\circ$ is a countable intersection of closed absolutely convex neighbourhoods of zero. Now the hypothesis that E is countably quasi-barrelled implies that B° is a neighbourhood of zero, i.e., B is equicontinuous.   □

We mention that, in Lemma 7.4, the property 'countably quasi-barrelled' is *equivalent* to the requirement that each $\beta(E', E)$-bounded countable union of equicontinuous subsets of E' is equicontinuous; see [Bou07a, Chap. IV, § 3,Proposition 1]. In fact, it is the latter condition that Grothendieck [Gro54, p. 63] takes into his definition of DF-spaces.

> **Theorem 7.5**
> Let E be a DF-space. Then E'_β is a Fréchet space.

Proof

By hypothesis, there exists a countable cobase $(B_n)_{n\in\mathbb{N}}$ of bounded subsets of E. Then (B_n°) is a neighbourhood base of zero in E'_β. This shows that E'_β is metrisable; let d be a translation invariant metric on E' inducing the topology $\beta(E', E)$ (see Proposition 2.17).

Let (y_n) be a Cauchy sequence in (E', d). Then for any $V \in \mathcal{U}_0(E'_\beta)$ there exists n_0 such that $y_n - y_m \in V$ for all $m, n \geqslant n_0$. This implies that the sequence (y_n) is bounded in E'_β. It also implies that for all $x \in E$ the sequence $\langle x, y_n \rangle$ is a Cauchy sequence in $\mathbb{K}$; hence

$$y(x) := \lim_{n\to\infty} \langle x, y_n \rangle \qquad (x \in E) \tag{7.3}$$

defines an element $y \in E^*$. Lemma 7.4 implies that $\{y_n\,;\; n \in \mathbb{N}\} = \bigcup_{n\in\mathbb{N}}\{y_n\}$ is equicontinuous, i.e., there exists $U \in \mathcal{U}_0(E)$ such that

$$\{y_n\,;\; n \in \mathbb{N}\} \subseteq U^\circ \subseteq U^\bullet = \{z \in E^*\,;\; |\langle x, z \rangle| \leqslant 1 \ (x \in U)\}.$$

Then (7.3) implies that $y \in U^\bullet$; hence Theorem 3.2 show that $y \in E'$. Finally, for each bounded set $B \subseteq E$, the sequence $\big(\langle \cdot, y_n \rangle|_B\big)_{n\in\mathbb{N}}$ is a Cauchy sequence with respect to the sup-norm, hence uniformly convergent to $\langle \cdot, y \rangle|_B$. This shows that $y_n \to y$ with respect to $\beta(E', E)$ as $n \to \infty$.   □

Combining the previous results we now obtain information on the bidual of metrisable locally convex spaces.

Corollary 7.6 *Let E be a metrisable locally convex space. Then its bidual E''_β ($:= (E'', \beta(E'', E'))$) is a Fréchet space, and E is isomorphic to a subspace of E''_β via the canonical embedding $\kappa: E \hookrightarrow E''$. In particular, if E is a Fréchet space, then E is isomophic to a closed subspace of E''_β.*

Proof

From Theorem 7.1 we know that E'_β is a DF-space, and then Theorem 7.5 implies that E''_β is a Fréchet space. As E is bornological, by Proposition 6.11, hence quasi-barrelled, the combination of Theorems 6.7 and 6.8 shows that κ is an isomorphism between E and $\kappa(E)$. If E is a Fréchet space, then $\kappa(E)$ is a complete, hence closed subspace of E''_β. More precisely, if d'' is a translation invariant metric on E'' inducing the topology $\beta(E'', E')$, then the restriction d of d'' to $\kappa(E)$ is a translation invariant metric on $\kappa(E)$ and $(\kappa(E), d)$ is complete, hence $\kappa(E)$ is closed in (E'', d''). □

As a preparation for the following example we need a property for polars which we did not use so far, and which we don't want to treat in the middle of the example. Let $\langle E, F \rangle$ be a dual pair, and let $\mathcal{A}$ be a collection of absolutely convex $\sigma(E, F)$-closed subsets of E. Then the bipolar theorem, Theorem 3.6, and Remark 3.3(c) imply

$$\left(\bigcap \mathcal{A}\right)^\circ = \left(\bigcap_{A \in \mathcal{A}} A^{\circ\circ}\right)^\circ = \left(\bigcup_{A \in \mathcal{A}} A^\circ\right)^{\circ\circ} = \overline{\mathrm{aco}}\left(\bigcup_{A \in \mathcal{A}} A^\circ\right), \qquad (7.4)$$

where the closure in the last term is with respect to $\sigma(F, E)$.

Examples 7.7
(a) Continuing the treatment of the space of rapidly decreasing sequences s and its dual t, we determine the bounded sets of s. Recall the norms p_k on s from Example 2.19(c). A set $B \subseteq s$ is bounded if and only if

$$\gamma_k := \sup_{x \in B} p_k(x) < \infty$$

for all $k \in \mathbb{N}_0$, and then $B \subseteq \bigcap_{k \in \mathbb{N}_0} B_{p_k}[0, \gamma_k]$. This implies that

$$\mathcal{B} := \left\{ \bigcap_{k \in \mathbb{N}_0} B_{p_k}[0, \gamma_k]; \ \gamma \in (0, \infty)^{\mathbb{N}_0} \right\}$$

is a cobase of bounded sets in s.

The 'dual norm' q_k on t to p_k in the dual pair $\langle s, t \rangle$ is given by

$$q_k(y) = \sup_{n \in \mathbb{N}} |y_n| n^{-k} \in [0, \infty] \qquad (y \in t),$$

and one obtains $B_{p_k}[0, c]^\circ = B_{q_k}[0, 1/c]$ for all $k \in \mathbb{N}_0$, $c > 0$, with the polar taken in $\langle s, t \rangle$.

This statement needs some explanation. For a dual pair $\langle E, F \rangle$ and a set $A \subseteq E$, we define $q_A : F \to [0, \infty]$ by

$$q_A(y) := \sup \left\{ |\langle x, y \rangle|; \ x \in A \right\} \qquad (y \in F).$$

(In Chapter 3, this definition was given for the case that A is $\sigma(E, F)$-bounded, yielding a semi-norm $q_A : F \to [0, \infty)$.) Extending our definition of balls we will use the notation

$$B_{q_A}[0, c] := \left\{ y \in F; \ q_A(y) \leqslant c \right\} \qquad (c > 0);$$

this still implies that $A^\circ = B_{q_A}[0, 1]$.

For the present case, the "norm" $q_k = q_A : t \to [0, \infty]$ corresponds to the set $A := B_{p_k}[0, 1] \subseteq s$. For $B = \bigcap_{k \in \mathbb{N}_0} B_{p_k}[0, \gamma_k] \in \mathcal{B}$ one then obtains, applying (7.4), a neighbourhood of zero for $\beta(t, s)$ by

$$B^\circ = \overline{\mathrm{aco}} \left(\bigcup_{k \in \mathbb{N}_0} B_{q_k}[0, 1/\gamma_k] \right),$$

and these sets constitute a neighbourhood base of zero when B runs through $\mathcal{B}$.

In Example 10.7 we will show that $\beta(t, s)$ is a locally convex inductive limit topology.

(b) Define

$$C^\infty[0, 1] := \left\{ f \in C^\infty(0, 1) ;\ f^{(n)} \text{ extends continuously to } [0, 1], \text{ for all } n \in \mathbb{N}_0 \right\},$$

with the topology τ generated by the semi-norms

$$p_n(f) := \|f^{(n)}\|_\infty \qquad (f \in C^\infty[0, 1],\ n \in \mathbb{N}_0).$$

It is standard to show that then $C^\infty[0, 1]$ is a Fréchet space.

For the description of the dual space of $C^\infty[0, 1]$ we first observe that the mapping

$$j : C^\infty[0, 1] \to E := C[0, 1]^{\mathbb{N}_0},\ f \mapsto \left(f^{(n)} \right)_{n \in \mathbb{N}_0}$$

is an isomorphic embedding, if E is provided with the topology generated by the sequence of semi-norms $(q_m)_{m \in \mathbb{N}_0}$,

$$q_m(g) := \|g_m\|_\infty \qquad \left(g = (g_n)_{n \in \mathbb{N}_0} \in C[0, 1]^{\mathbb{N}_0},\ m \in \mathbb{N}_0 \right).$$

As for $\mathbb{K}^{\mathbb{N}}$ – see Examples 1.10 and 2.19(a) –, one shows that E is also a Fréchet space and that the dual space of E is given by

$$E' = \bigoplus_{n \in \mathbb{N}_0} \mathcal{M}[0, 1] := \left\{ (\mu_n)_{n \in \mathbb{N}_0} \in \mathcal{M}[0, 1]^{\mathbb{N}_0} ;\ \exists m \in \mathbb{N}_0 : \mu_n = 0\ (n \geqslant m) \right\},$$

where $\mathcal{M}[0, 1]$ denotes the space of signed Borel measures on $[0, 1]$. (Here one also has to use that $C[0, 1]'$ is isomorphic to $\mathcal{M}[0, 1]$, by the Riesz–Markov representation theorem; see [Rud87, Theorem 2.14].)

Now let $\eta \in C^\infty[0, 1]'$. Then $\eta \circ j^{-1} \in j(C^\infty[0, 1])'$, and applying Corollary 2.16 we obtain $\tilde{\eta} \in E'$ such that $\tilde{\eta}|_{j(C^\infty[0,1])} = \eta \circ j^{-1}$.

In order to express this differently we define the dual pair $\langle C^\infty[0, 1]), E' \rangle$, with duality bracket

$$\langle f, \tilde{\eta} \rangle := \langle j(f), \tilde{\eta} \rangle_{E, E'} = \sum_{n \in \mathbb{N}_0} \int_{[0, 1]} f^{(n)}\, d\mu_n,$$

where $\tilde{\eta} = (\mu_n)_{n \in \mathbb{N}_0}$. In this setting we have shown above that $C^\infty[0, 1]' = b_2(E')$, where b_2 is the mapping $b_2 \colon E' \to C^\infty[0, 1]^*$ defined in Chapter 1, for the dual pair $\langle C^\infty[0, 1]), E'\rangle$. Not unexpectedly, b_2 is far from being injective: For instance, if $0 \neq \varphi \in C_c^1(0, 1)$ and we define $\tilde{\eta} = (\mu_n)$ by $\mu_0 := -\varphi'\lambda$, $\mu_1 := \varphi\lambda$ (where λ denotes the Lebesgue measure), $\mu_n := 0$ $(n \geqslant 2)$, then $\tilde{\eta} \neq 0$, but

$$\langle f, \tilde{\eta} \rangle = \langle j(f), \tilde{\eta} \rangle_{E,E'} = \int_0^1 f(-\varphi')\,dx + \int_0^1 f'\varphi\,dx = 0$$

for all $f \in C^\infty[0, 1]$. As a consequence, defining $(E')_0 := \ker b_2$, we obtain $C^\infty[0, 1]'$ as the quotient space $E'/(E')_0$. △

In the previous example we have seen a natural description of the dual of a Fréchet space as a quotient space. Further instances of this case are Example 8.4(b) or the space treated in Chapter 16. We take this as a motivation for a short interlude on topologies on quotient spaces. We start with preparations concerning the final topology for topological spaces.

Remarks 7.8 (a) Let (X, τ) be a topological space, Y a set, and $f \colon X \to Y$. On Y we define the **final topology** by

$$\sigma := \{V \subseteq Y;\ f^{-1}(V)\ \text{open}\}.$$

It is easy to see that indeed σ is a topology, clearly the finest topology such that $f \colon X \to Y$ is τ-σ-continuous.

Suppose additionally that f is surjective. Then one has $\sigma \subseteq \{f(U);\ U \in \tau\}$. Indeed, for $V \in \sigma$ one obtains $f^{-1}(V) \in \tau$ and $V = f(f^{-1}(V))$.

The mapping f defines an equivalence relation on X, with equivalence classes given by $f^{-1}(y)$, for $y \in Y$. Assume that for all open sets $U \subseteq X$ the union $f^{-1}(f(U))$ of the equivalence classes of elements belonging to U is open. Then $f(U) \in \sigma$ by definition, and one concludes that $\sigma = \{f(U);\ U \in \tau\}$.

In particular, the mapping f is open, i.e., $f(U)$ is open for all open sets U.

(b) Let X, Y, Z be topological spaces, let $f \colon X \to Y$ be continuous, open and surjective, and let $g \colon Y \to Z$.

Then $g \circ f$ is continuous if and only if g is continuous. Indeed, assume that $g \circ f$ is continuous, and let $W \subseteq Z$ be open. Then $f^{-1}(g^{-1}(W)) = (g \circ f)^{-1}(W)$ is open, hence $g^{-1}(W) = f(f^{-1}(g^{-1}(W)))$ is open; so g is continuous. The reverse implication is trivial. △

The following theorem is the basic result on quotient spaces and their topologies. We state it without explicitly mentioning the quotient, but in the given context one can interpret F as the quotient of E by $\ker q$.

Theorem 7.9
Let (E, τ) be a topological vector space, F a vector space, $q \colon E \to F$ linear and surjective, and let σ be the final topology on F with respect to q.
  (a) Then q is a continuous and open mapping, and (F, σ) is a topological vector space.
  (b) The topology σ is Hausdorff if and only if $\ker q$ is closed.

Proof

(a) Let $E_0 := \ker q$. Then for each $A \subseteq E$ the set $q^{-1}(q(A))$ is given by $A + E_0$. This implies that for open $U \subseteq E$ the set $q^{-1}(q(U)) = U + E_0 = \bigcup_{x \in E_0}(x + U)$ is open. Now Remark 7.8(a) implies that q is open.

Let $a \colon E \times E \to E$ be addition in E, and denote by $\tilde{a} \colon F \times F \to F$ addition in F; then

$$q \circ a = \tilde{a} \circ (q \times q).$$

It is easy to see that $q \times q \colon E \times E \to F \times F$ is open. (Use that $(q \times q)(U_1 \times U_2) = q(U_1) \times q(U_2)$ is open in $F \times F$ for all open sets $U_1, U_2 \subseteq E$, and recall that the sets $U_1 \times U_2$ constitute a base of the product topology.). Therefore, the continuity of $q \circ a$ and Remark 7.8(b) imply that $\tilde{a}$ is continuous. The continuity of scalar multiplication in F can be proved analogously.

(b) The topology σ on F is Hausdorff if and only if $F \setminus \{0\} = q(E \setminus \ker q)$ is open, and – because q is open – the latter holds if and only if $E \setminus \ker q$ is open. □

Remark 7.10 For many Banach spaces one knows rather well, or even explicitly, the dual space. The corresponding issue for Fréchet spaces is more involved. One of the questions is already what is meant by "knowing the dual space". The best answer would always be: having an expression in terms of known spaces. Another answer would be to know that the dual space is isomorphic to a known space.

Except for the space s, we will not look further at this problem, but rather refer to the treatise of Meise and Vogt [MeVo97], where a whole chapter is devoted to an exhaustive treatment of this issue. △

The next (and final) topic of this chapter is to derive an alternative description of Fréchet spaces. To formulate this description, we define a topological space (X, τ) to be **completely metrisable** if there exists a metric d inducing τ and such that (X, d) is a complete metric space. Clearly, each Fréchet space is completely metrisable, even by a translation invariant metric. It should come as a surprise that the translation invariance is not really needed.

Theorem 7.11
Let E be a completely metrisable locally convex space. Then E is a Fréchet space.

The proof depends on Baire's theorem, see Appendix B, and requires some further preparation. A subset A of a topological space X is called a $\boldsymbol{G_\delta}$-**set** if it is a countable intersection of open subsets of X.

Proposition 7.12 (Sierpiński) *Let (X, d) be a metric space, $Y \subseteq X$, and let e be a metric on Y such that (Y, e) is a complete metric space and such that d and e induce the same topology on Y. Then Y is a G_δ-subset of X.*

Proof

(i) Let $n \in \mathbb{N}$. Then for each $y \in Y$ there exists $r_n(y) \in (0, 1/n)$ such that $B_d(y, r_n(y)) \cap Y \subseteq B_e(y, 1/n)$. (The ball $B_e(y, 1/n)$ is taken in Y, where the metric e is defined.) Then $G_n := \bigcup_{y \in Y} B_d(y, r_n(y))$ is an open subset of (X, d); hence $\Gamma := \bigcap_{n \in \mathbb{N}} G_n$ is a G_δ-subset of (X, d), and clearly $Y \subseteq \Gamma$.

(ii) Now let us show that $\Gamma \subseteq Y$. Let $x \in \Gamma$. For every $n \in \mathbb{N}$ there exists $y_n \in Y$ such that $x \in B_d(y_n, r_n(y_n))$. This implies that $y_n \to x$ in (X, d).

Let $\varepsilon > 0$; then there exists $n \in \mathbb{N}$ such that $2/n \leqslant \varepsilon$. As $d(x, y_n) < r_n(y_n)$, we find $m \in \mathbb{N}$ such that

$$d(x, y_n) + 1/m < r_n(y_n).$$

For $k \in \mathbb{N}, k \geqslant m$ we then obtain

$$d(y_k, y_n) \leqslant d(y_k, x) + d(x, y_n) < 1/k + d(x, y_n) < r_n(y_n),$$

hence

$$y_k \in B_d(y_n, r_n(y_n)) \cap Y \subseteq B_e(y_n, 1/n).$$

For $j, k \in \mathbb{N}, j, k \geqslant m$ we conclude that $e(y_j, y_k) \leqslant e(y_j, y_n) + e(y_n, y_j) < 2/n \leqslant \varepsilon$.

So we have shown that (y_n) is a Cauchy sequence in (Y, e), and the completeness of (Y, e) implies that there exists $y \in Y$ such that $y_n \to y$ in (Y, e). This shows that $x = d\text{-}\lim y_n = e\text{-}\lim y_n = y \in Y$. $\qquad \square$

Remarks 7.13 (a) An instructive easy example for the situation in Proposition 7.12 is the choice $X := \mathbb{R}$ with the distance metric d, and $Y := (0, 1)$ with the metric $e(x, y) := |g(x) - g(y)|$, for any continuous, strictly monotonically increasing function $g \colon (0, 1) \to \mathbb{R}$ satisfying $\lim_{x \to 0+} g(x) = -\infty$, $\lim_{x \to 1-} g(x) = \infty$. In this case the G_δ property of $Y = (0, 1)$ is trivially satisfied.

(b) In order to indicate more sophisticated examples we mention that it can be shown that on any G_δ-subset Y of a complete metric space X one can find a complete metric on Y that is topologically equivalent to the restriction of the original metric to Y; see [Wil70, Theorem 24.12]. △

Another tool we will need is a consequence of the duality of Fréchet spaces treated above.

Proposition 7.14 *Let E be a metrisable locally convex space. Then there exist a Fréchet space $\tilde{E}$ and an embedding $\kappa \colon E \hookrightarrow \tilde{E}$ such that $\kappa \colon E \to \kappa(E)$ is an isomorphism (of locally convex spaces), and $\kappa(E)$ is dense in $\tilde{E}$.*

Proof
Let $\kappa \colon E \to E''$ as in Corollary 7.6, and define $\tilde{E} := \overline{\kappa(E)}^{E''_\beta}$. Then $\tilde{E}$ is a Fréchet space with the asserted properties. □

Proof of Theorem 7.11
Let e be a metric on E inducing the topology of E, and such that (E, e) is complete. As E is metrisable, there exists a Fréchet space $\tilde{E} \supseteq E$ as in Proposition 7.14, with a translation invariant metric d inducing the topology of $\tilde{E}$. We will show that in fact $E = \tilde{E}$, which then proves the assertion, because on E the metrics d and e induce the same topology.

To obtain a contradiction, suppose that $E \subset \tilde{E}$. For the following discussion we refer to Appendix B, concerning the terminology and the results. Applying Proposition 7.12 we conclude that E is a dense G_δ-set, hence a residual set in $(\tilde{E}, d)$. This implies that $\tilde{E} \setminus E$ is a meagre subset of $(\tilde{E}, d)$. Let $x \in \tilde{E} \setminus E$. Then $h_x \colon \tilde{E} \to \tilde{E}$, $y \mapsto y + x$ is a homeomorphism mapping E to a subset of $\tilde{E} \setminus E$. (Indeed, $y \in E$ together with $y + x \in E$ would imply $x \in E$.) This shows that $h_x(E)$, and hence E is a meagre subset of $(\tilde{E}, d)$. This is a contradiction, because in the Baire space $(\tilde{E}, d)$ a set cannot be residual and meagre simultaneously; see Proposition B.2. □

Notes We have adopted the notion 'countably quasi-barrelled space' from Khalleelula [Kha82]; in [Bou07a, Chap. IV, § 3] such spaces are called 'semi-barrelled'. Also, we have adopted the notion 'cobase of bounded sets' from [Wil78, Section 1–6, Exercise 106]. The terminology 'DF-space' (French: 'espace (DF)') was coined by Grothendieck [Gro54]. Obviously, 'DF' stands for '**d**ual **F**réchet', and indeed the DF-space properties are essential ingredients of duals and preduals of Fréchet spaces; see Theorems 7.1 and 7.5. Grothendieck [Gro54, p. 64] comments that the fact that dual spaces of DF-spaces are Fréchet spaces 'justifie notre terminologie'. Our treatment follows mainly [Bou07a, Chap. IV, § 3] and [MeVo97, Section 25].

Motivated by Example 7.7(b), we have included some information on the topology of quotient spaces.

The last part of this chapter describes Fréchet spaces as 'completely metrisable locally convex spaces'; see Theorem 7.11. Our source for this – seemingly not widely known – fact is [Wil78, Section 4–5, Exercise 104]. The essential auxiliary fact needed in the proof is Proposition 7.12, due to Sierpiński [Sie28].

Reflexivity

We start by discussing semi-reflexivity and Montel spaces and present a number of examples of function spaces. At the end we present duality properties for reflexive spaces and Montel spaces.

We recall from Chapter 3 that a locally convex space E is called semi-reflexive if it is Hausdorff and the canonical embedding $\kappa\colon E \hookrightarrow E''$ is surjective. E is called reflexive if additionally κ is continuous, where the image space is equipped with the strong topology.

From Theorems 6.7 and 6.8 we know that (E, τ) is reflexive if and only if E is semi-reflexive and quasi-barrelled, or equivalently (because always $\tau \subseteq \beta(E, E')$, by Proposition 6.4) if and only if E is semi-reflexive, and $\tau = \beta(E, E')$, or equivalently (by Theorem 6.14), if and only if E is semi-reflexive and barrelled.

This is the reason why in the following we will mainly discuss semi-reflexivity.

Theorem 8.1

Let E be a Hausdorff locally convex space. Then E is semi-reflexive if and only if every bounded set in E is weakly relatively compact.

Proof

For the necessity we note that semi-reflexivity implies that $\beta(E', E) = \mu(E', E)$. Therefore, if $A \subseteq E$ is bounded, then A° is a $\mu(E', E)$-neighbourhood of zero, and there exists a $\sigma(E, E')$-compact barrel $C \subseteq E$ such that $A^\circ \supseteq C^\circ$. Then $A \subseteq A^{\circ\circ} \subseteq C^{\circ\circ} = C$.

For the sufficieny we note that the condition implies that $\beta(E', E) = \mu(E', E)$, which in turn implies that $(E', \beta(E', E))' = (E', \mu(E', E))' = E$. $\qquad\square$

Remark 8.2 Note that the condition in Theorem 8.1 is a generalisation of the known criterion for the reflexivity of Banach spaces. $\qquad\triangle$

© Springer Nature Switzerland AG 2020
J. Voigt, *A Course on Topological Vector Spaces*, Compact Textbooks in Mathematics,
https://doi.org/10.1007/978-3-030-32945-7_8

A **semi-Montel space** is a Hausdorff locally convex space in which every bounded set is relatively compact. (This terminology reminds of Montel's theorem from complex analysis; see Example 8.4(d).) A **Montel space** is a quasi-barrelled semi-Montel space.

Corollary 8.3 *If E is a semi-Montel space, then E is semi-reflexive. If E is a Montel space, then E is reflexive.*

Proof
This is obvious from Theorem 8.1.   □

For use in the following example (b) we mention the notation $C_0(\Omega)$, for the space of continuous functions 'vanishing at ∞', on a Hausdorff locally compact space Ω:

$$C_0(\Omega) := \big\{ f \in C(\Omega) \,;\, \forall \varepsilon > 0 \,\exists\, K \subseteq \Omega \text{ compact} \colon |f(x)| < \varepsilon \ (x \in \Omega \setminus K) \big\}.$$

For a function $f \in C(\Omega)$, the **support** is defined by $\operatorname{spt} f := \overline{\{ x \in \Omega \,;\, f(x) \neq 0 \}}$.

Examples 8.4
(a) The space s of rapidly decreasing sequences is a **Fréchet–Montel space**, i.e., a Fréchet space which also is semi-Montel (hence Montel, because Fréchet spaces are barrelled). Indeed, if $(x^k)_{k \in \mathbb{N}}$ is a bounded sequence in s, then one can choose a subsequence converging in each coordinate, and it is easy to show that this subsequence is convergent in s. Hence s is reflexive.

(b) Let $\Omega \subseteq \mathbb{R}^n$ be open and bounded. Then

$$C_0^\infty(\Omega) := \big\{ f \in C^\infty(\Omega) \,;\, \partial^\alpha f \in C_0(\Omega) \ (\alpha \in \mathbb{N}_0^n) \big\},$$

with norms

$$p_m(f) := \max \big\{ \|\partial^\alpha f\|_\infty \,;\, |\alpha| \leqslant m \big\} \quad (m \in \mathbb{N}_0, \ f \in C_0^\infty(\Omega)),$$

is a Fréchet–Montel space, therefore reflexive.

Indeed, $C_0^\infty(\Omega)$ is a Fréchet space. Also, every bounded set is relatively compact because of the Arzelà–Ascoli theorem, and therefore the space is semi-Montel.

A partial description of the dual is given as follows. If $\eta \in C_0^\infty(\Omega)'$, then there exist $m \in \mathbb{N}_0$ and $c \geqslant 0$ such that $|\eta(f)| \leqslant c\,p_m(f)$ ($f \in C_0^\infty(\Omega)$). The mapping

$$\Phi \colon (C_0^\infty(\Omega), p_m) \to C_0(\Omega)^{\{\alpha; \, |\alpha| \leqslant m\}}, \quad f \mapsto \big(\partial^\alpha f\big)_{|\alpha| \leqslant m},$$

is linear and isometric, and therefore the Hahn–Banach theorem implies that there exists $\hat\eta \in \big(C_0(\Omega)^{\{\alpha; \, |\alpha| \leqslant m\}} \big)'$ such that $\hat\eta \circ \Phi = \eta$. The Riesz–Markov theorem (see [Rud87,

Theorem 2.14]) implies that there exists a family $\left(\mu_\alpha\right)_{|\alpha|\leqslant m}$ of finite Borel measures on Ω such that

$$\hat{\eta}(g) = \sum_{|\alpha|\leqslant m} \int g_\alpha \, d\mu_\alpha \quad \left(g = \left(g_\alpha\right)_{|\alpha|\leqslant m} \in C_0(\Omega)^{\{\alpha;\ |\alpha|\leqslant m\}}\right).$$

For $f \in C_0^\infty(\Omega)$ this means that

$$\eta(f) = \sum_{|\alpha|\leqslant m} \int \partial^\alpha f \, d\mu_\alpha = \left(\sum_{|\alpha|\leqslant m} (-1)^{|\alpha|} \partial^\alpha \mu_\alpha\right) f,$$

where the derivatives of the measures should be interpreted in the sense of distributions. (Strictly speaking, the last formula would only be valid for $f \in C_c^\infty(\Omega)$, but the distributions can be extended by continuity to $f \in C_0^\infty(\Omega)$.)

(c) Let $\Omega \subseteq \mathbb{R}^n$ be open. Then $\mathcal{E}(\Omega) := C^\infty(\Omega)$, with semi-norms

$$p_{K,m}(f) := \max\left\{\|\partial^\alpha f\|_K\, ; \ |\alpha| \leqslant m\right\} \quad (K \subseteq \Omega \text{ compact}, \ m \in \mathbb{N}_0, \ f \in \mathcal{E}(\Omega))$$

(where $\|\cdot\|_K$ denotes the sup-norm on K) is a Fréchet–Montel space, in particular reflexive.

Let $\left(\Omega_k\right)_{k\in\mathbb{N}}$ be a **standard exhaustion** of Ω, i.e., Ω_k is open, relatively compact in Ω_{k+1} ($k \in \mathbb{N}$), and $\bigcup_{k\in\mathbb{N}} \Omega_k = \Omega$. Define $K_k := \overline{\Omega_k}$ ($k \in \mathbb{N}$). Then any compact subset of Ω is contained in some K_k; hence the topology of $\mathcal{E}(\Omega)$ is generated by the set $\left\{p_{K_k,m}\, ;\ k \in \mathbb{N}, \ m \in \mathbb{N}_0\right\}$; therefore $\mathcal{E}(\Omega)$ is metrisable, and also it is complete. (Note that, even though we use the standard exhaustion for the proof of the above properties, the topology does not depend on the choice of the exhaustion.)

Next we sketch why $\mathcal{E}(\Omega)$ is semi-Montel. As an intermediate step let $k \in \mathbb{N}_0$, and let (f_j) be a sequence in $\mathcal{E}(\Omega)$, $\sup_j \left\{\|\partial_l f_j\|_{K_{k+1}}\, ;\ 1 \leqslant l \leqslant n\right\} < \infty$. Then the sequence (f_j) is bounded on K_{k+1} and equicontinuous on K_k, and by the Arzelà–Ascoli theorem there exists a $\|\cdot\|_{K_k}$-Cauchy subsequence. Now let (f_j) be a bounded sequence in $\mathcal{E}(\Omega)$. This means that $\sup_j p_{K,m}(f_j) < \infty$ for all compact $K \subseteq \Omega$, $m \in \mathbb{N}_0$. Applying the previous remark and a suitable diagonal procedure one obtains a subsequence which is a $p_{K,m}$-Cauchy sequence for all compact $K \subseteq \Omega$, $m \in \mathbb{N}_0$, i.e., a Cauchy sequence, and therefore convergent in $\mathcal{E}(\Omega)$.

(d) Let $\Omega \subseteq \mathbb{C}$ be open, $\mathcal{H}(\Omega) := \left\{f\colon \Omega \to \mathbb{C}\, ;\ f \text{ holomorphic}\right\}$, with semi-norms

$$p_K(f) := \|f\|_K \quad (f \in \mathcal{H}(\Omega), \ K \subseteq \Omega \text{ compact}).$$

Then $\mathcal{H}(\Omega)$ is a Fréchet–Montel space, therefore reflexive.

The Montel property of $\mathcal{H}(\Omega)$ is just **Montel's theorem**, and for completeness we recall its proof. Let $H \subseteq \mathcal{H}(\Omega)$ be a bounded set. Let (Ω_n) be a standard exhaustion of Ω, and for $n \in \mathbb{N}$ let $K_n := \overline{\Omega_n}$. For all $n \in \mathbb{N}$ one has

$$C_n := \sup\left\{\|f\|_{K_n}\, ;\ f \in H\right\} < \infty,$$

and there exists $r_n > 0$ such that $K_n + B_{\mathbb{C}}[0, r_n] \subseteq K_{n+1}$. Then Cauchy's integral formula for the derivative,

$$f'(z) = \frac{1}{2\pi i} \int_{\partial B(z,r)} \frac{f(\zeta)}{(\zeta - z)^2} \, d\zeta,$$

implies that $|f'(z)| \leqslant C_{n+1}(r_n/2)^{-2}$ for all $z \in K_n + B_{\mathbb{C}}(0, r_n/2)$, $f \in H$, and this estimate shows that $H_{K_n} := \{ f|_{K_n} ; \ f \in H \}$ is equicontinuous. From the Arzelà–Ascoli theorem we conclude that H_{K_n} is a relatively compact subset of $C(K_n)$.

Now, starting with a sequence (f_k) in H we can choose a subsequence $(f_{k_j})_{j \in \mathbb{N}}$ such that $(f_{k_j}|_{K_n})_{j \in \mathbb{N}}$ converges in $C(K_n)$ for all $n \in \mathbb{N}$, i.e., $(f_{k_j})_{j \in \mathbb{N}}$ is convergent in $C(\Omega)$. This shows that H is relatively sequentially compact in the metric space $\mathcal{H}(\Omega)$, hence relatively compact.

(e) Let $\Omega \subseteq \mathbb{R}^n$ be open,

$$H(\Omega) := \{ f \in C^2(\Omega) ; \ f \text{ harmonic} \},$$

with semi-norms

$$p_K(f) := \| f \|_K \quad (K \subseteq \Omega \text{ compact}, \ f \in H(\Omega)).$$

We recall that **harmonic** means that $\Delta f = \sum_{j=1}^n \partial_j^2 f = 0$. We will explain that then $H(\Omega)$ is a Fréchet–Montel space.

(i) Let $P := \sum_{|\alpha| \leqslant m} a_\alpha \partial^\alpha$ be a partial differential operator with constant coefficients. Then it is easy to see that the space

$$E_P(\Omega) := \{ f \in \mathcal{E}(\Omega) ; \ Pf = 0 \}$$

is a closed subspace of $\mathcal{E}(\Omega)$, therefore a Fréchet–Montel space; see Theorem 8.8(b) below. In the following we will sketch why $H(\Omega) = E_\Delta(\Omega)$.

(ii) We recall that harmonic functions f have the mean value property, i.e., if $x \in \Omega$, $r > 0$ are such that $B[x, r] \subseteq \Omega$, then

$$f(x) = \frac{1}{\sigma_{n-1}} \int_{S_{n-1}} f(x + r\xi) \, dS(\xi).$$

We refer to [Eva98, Section 2.2.2, Theorem 2] (or any other textbook on partial differential equations) for this property.

(iii) Let $(\Omega_k)_{k \in \mathbb{N}}$ be a standard exhaustion of Ω, $K_k := \overline{\Omega_k}$, $d_k := \text{dist}(K_k, \Omega \setminus \Omega_{k+1})$, and let $\rho_k \in C_c^\infty(\mathbb{R}^n)$, $\rho_k \geqslant 0$, $\text{spt } \rho_k \subseteq B(0, d_k)$, $\int \rho_k(x) \, dx = 1$, $\rho_k(x) = \rho_k(y)$ if $|x| = |y|$ $(k \in \mathbb{N})$. Then, for $f \in H(\Omega)$, the convolution $\rho_k * f$,

$$\rho_k * f(x) := \int_{\Omega_{k+1}} \rho_k(x - y) f(y) \, dy,$$

is defined for $x \in \Omega_k$, and in fact is equal to $f(x)$, because of the mean value property of f. Differentiating under the integral sign, one concludes that f is infinitely differentiable on Ω_k, and that

$$\partial^\alpha f(x) = \int_{\Omega_{k+1}} \partial^\alpha \rho_k(x - y) f(y)\, dy \quad (x \in \Omega_k, \ \alpha \in \mathbb{N}_0^n).$$

(iv) From (iii) it follows that, for $k \in \mathbb{N}$, $\alpha \in \mathbb{N}_0^n$ there exists a constant $c_{k,\alpha}$ such that

$$\|\partial^\alpha f\|_{K_k} \leqslant c_{k,\alpha} \|f\|_{K_{k+1}} \quad (f \in H(\Omega)).$$

This shows that the topology on $H(\Omega)$ defined above is the topology induced by $\mathcal{E}(\Omega)$. Therefore the assertion follows from (i).

(f) The **Schwartz space** $\mathcal{S}(\mathbb{R}^n)$, also called the space of **rapidly decreasing functions**, is defined by

$$\mathcal{S}(\mathbb{R}^n) := \left\{ f \in C^\infty(\mathbb{R}^n); \ x \mapsto (1 + |x|^2)^m \partial^\alpha f(x) \text{ bounded } (m \in \mathbb{N}_0, \ \alpha \in \mathbb{N}_0^n) \right\},$$

with norms

$$p_{m,k}(f) := \max \left\{ (1 + |x|^2)^m |\partial^\alpha f(x)|; \ x \in \mathbb{R}^n, \ |\alpha| \leqslant k \right\} \ (m, k \in \mathbb{N}_0, \ f \in \mathcal{S}(\mathbb{R}^n)).$$

It is standard to show that $\mathcal{S}(\mathbb{R}^n)$ is a Fréchet space. Next we show that $\mathcal{S}(\mathbb{R}^n)$ is a Montel space.

Let $m \in \mathbb{N}_0$, (f_k) a sequence with $M := \sup_k p_{m+1,m+1}(f_k) < \infty$. We show that then there exists a $p_{m,m}$-Cauchy subsequence. Let $\varepsilon > 0$; choose $R > 0$ such that $\frac{M}{1+R^2} < \varepsilon$. Then

$$\sup_k \left\{ (1 + |x|^2)^m |\partial^\alpha f_k(x)|; \ |x| \geqslant R, \ |\alpha| \leqslant m \right\} \leqslant \varepsilon.$$

For $|\alpha| \leqslant m$ the set $\{\partial^\alpha f_k; \ k \in \mathbb{N}\}$ is $\|\cdot\|_\infty$-bounded and equicontinuous on $B[0, R]$, and therefore, by the Arzelà–Ascoli theorem, there exists a subsequence $(f_{k_j})_j$ such that $(\partial^\alpha f_{k_j})_j$ is $\|\cdot\|_\infty$-convergent on $B[0, R]$, for all $|\alpha| \leqslant m$. Repeating this argument for smaller and smaller ε and choosing suitable subsequences, we obtain a $p_{m,m}$-Cauchy subsequence.

If (f_k) is a bounded sequence in $\mathcal{S}(\mathbb{R}^n)$, then the previous procedure can be carried out for arbitrary $m \in \mathbb{N}$, yielding a Cauchy sequence in $\mathcal{S}(\mathbb{R}^n)$.

We mention the remarkable fact that $\mathcal{S}(\mathbb{R})$ is isomorphic to the space s; see [MeVo97, Example 29.5(2)]. An analogous result for $\mathcal{S}(\mathbb{R}^n)$ is presented in [ReSi80, Theorem V.13]. $\triangle$

After these examples we come back to some further theory.

Theorem 8.5

Let E be a reflexive Hausdorff locally convex space. Then the space $(E', \beta(E', E))$ is reflexive.

Proof

Let τ be the topology of E. By hypothesis and Theorem 6.8, $(E, \tau) = (E'', \beta(E'', E'))$. Therefore $E''' = (E'', \beta(E'', E'))' = (E, \tau)' = E'$, with $\beta(E''', E'') = \beta(E', E)$.   □

Theorem 8.6

Let E be a Montel space. Then $(E', \beta(E', E))$ is a Montel space.

For the proof we need a preparation. Let E be a topological vector space. We define the topology τ_c on E' to be the **topology of compact convergence**, i.e., the polar topology $\tau_{\mathcal{M}_c}$ corresponding to the collection $\mathcal{M}_c$ of compact subsets of E.

The fact proved next is, in principle, a property of a uniformly equicontinuous set of functions on a *uniform space*; topological vector spaces are special uniform spaces. In fact, part of the proof is just a generalised version of the proof of the following standard property: If B is an equicontinuous set of functions on a compact metric space A, and $f \in B$, $\varepsilon > 0$, then there exists a finite set $F \subseteq A$ such that

$$\left\{ g \in B; \ \sup_{x \in F} |g(x) - f(x)| \leqslant \varepsilon/3 \right\} \subseteq \left\{ g \in B; \ \|g - f\|_\infty \leqslant \varepsilon \right\}.$$

Proposition 8.7 *Let E be a topological vector space, and let $B \subseteq E'$ be equicontinuous. Then $\tau_c \cap B = \sigma(E', E) \cap B$.*

Proof

The inclusion '$\supseteq$' follows from $\tau_c \supseteq \sigma(E', E)$. For '$\subseteq$' it is sufficient to show: For $y_0 \in B$ and compact A, there exists a finite set $F \subseteq E$ such that

$$\left\{ y \in B; \ \sup_{x \in F} |\langle x, y - y_0 \rangle| \leqslant 1/3 \right\} \subseteq \left\{ y \in B; \ \sup_{x \in A} |\langle x, y - y_0 \rangle| \leqslant 1 \right\}.$$

(This property expresses that each τ_c-neighbourhood in B of y_0 contains a suitable $\sigma(E', E)$-neighbourhood in B of y_0.) As B is equicontinuous, there exists a balanced $U \in \mathcal{U}_0$ such that

$$\sup_{x \in U, \, y \in B} |\langle x, y \rangle| \leqslant 1/3.$$

Due to the compactness of A, there exists a finite set $F \subseteq A$ such that $A \subseteq F + U$. Now let $y \in B$ be such that $\sup_{\tilde{x} \in F} |\langle \tilde{x}, y - y_0 \rangle| \leqslant 1/3$. For $x \in A$ there exists $\tilde{x} \in F$ such that $x - \tilde{x} \in U$, which implies that

$$|\langle x, y - y_0 \rangle| \leqslant |\langle x - \tilde{x}, y \rangle| + |\langle \tilde{x}, y - y_0 \rangle| + |\langle \tilde{x} - x, y_0 \rangle| \leqslant 1;$$

hence $\sup_{x \in A} |\langle x, y - y_0 \rangle| \leqslant 1$. □

Proof of Theorem 8.6

The space $(E', \beta(E', E))$ is reflexive, by Corollary 8.3, therefore barrelled. Let $B \subseteq E'$ be $\beta(E', E)$-bounded, convex and closed. Theorem 6.8 implies that B is equicontinuous, therefore $\sigma(E', E)$-compact (by the Alaoglu–Bourbaki theorem). Now Proposition 8.7 implies that B is τ_c-compact. Since E is a Montel space, $\tau_c \cap E' = \beta(E', E)$, and therefore B is $\beta(E', E)$-compact. □

Theorem 8.8
Let E be a locally convex space, $F \subseteq E$ a closed subspace. Then:
(a) *If E is semi-reflexive, then F is semi-reflexive.*
(b) *If E is a semi-Montel space, then F is a semi-Montel space.*

Proof
(a) is a consequence of Theorem 8.1, because $\sigma(F, F') = \sigma(E, E') \cap F$ (recall Corollary 2.16).

(b) is obvious. □

Remark 8.9 The analogue of Theorem 8.8 with 'reflexive' instead of 'semi-reflexive' or 'Montel' instead of 'semi-Montel' does not hold. There even exists a Montel space with a non-reflexive closed subspace. We refer to [Sch71, Chap. IV, Exercises 19, 20] for an example. △

Notes The author was not able to trace the origins of (semi-)reflexivity and the (semi-)Montel property in locally convex spaces. The examples are standard in analysis. The isomorphy of $\mathcal{S}(\mathbb{R})$ and s, mentioned in Example 8.4(f) is due to Simon [Sim71, Theorem 1]. Theorem 8.6 can be found in [Köt66, VI, § 27.2], [Sch71, Chap. IV, § 5.9].

Completeness

Completeness is a property of a topological vector space as a 'uniform space'. We do not explicitly use uniform spaces but mention that the linear structure allows to define neighbourhoods of 'uniform size' for all $x \in E$ by taking the translates $x + U$ for $U \in \mathcal{U}_0(E)$. This allows to introduce the notion of Cauchy filters, and completeness requires Cauchy filters to be convergent.

After some discussion on completeness and quasi-completeness, we come to Grothendieck's description of the completion of a locally convex space, Corollary 9.16, as the main result of this chapter.

Let E be a topological vector space, $A \subseteq E$. A filter $\mathcal{F}$ in A is called a **Cauchy filter** if for every $U \in \mathcal{U}_0(E)$ there exists $B \in \mathcal{F}$ such that $B - B \subseteq U$.

The set $A \subseteq E$ is called **complete** if every Cauchy filter in A is convergent to an element of A, and A is called **sequentially complete** if every Cauchy sequence in A is convergent to an element of A. A sequence (x_n) in E is called a **Cauchy sequence** if the elementary filter generated by the sequence is a Cauchy filter, i.e., if for each neighbourhood of zero U there exists $n_0 \in \mathbb{N}$ such that $x_n - x_m \in U$ for all $m, n \geqslant n_0$.

The space E is called **quasi-complete** if every closed bounded subset of E is complete.

Remarks 9.1 (a) If $\mathcal{F}$ is a filter in A, $\mathcal{F}$ convergent to $x \in A$, then $\mathcal{F}$ is a Cauchy filter. (Let U be a neighbourhood of zero. Then there exists a neighbourhood of zero V such that $V - V \subseteq U$. Then $(x + V) \cap A \in \mathcal{F}$, by hypothesis, and one obtains $((x + V) \cap A) - ((x + V) \cap A) \subseteq V - V \subseteq U$.)

(b) Let $\mathcal{F}$ be a Cauchy filter in A, and let $x \in A$ be a cluster point of $\mathcal{F}$. Then $\mathcal{F} \to x$. (Let U be a neighbourhood of zero, V a neighbourhood of zero with $V + V \subseteq U$, $B \in \mathcal{F}$ with $B - B \subseteq V$ (in particular, $B \subseteq b + V$ for all $b \in B$). Then $B \cap (x + V) \neq \varnothing$, and therefore $B \subseteq B \cap (x + V) + V \subseteq x + V + V \subseteq x + U$. This shows that $\mathcal{F} \to x$.)

(c) If E is Hausdorff and A is complete, then A is closed. (For $x \in \overline{A}$ there exists a filter $\mathcal{F}$ in A with $\mathcal{F} \to x$. Then $\mathcal{F}$ is a Cauchy filter, which is convergent in A. Then $x \in A$, as the limit is unique.)

© Springer Nature Switzerland AG 2020
J. Voigt, *A Course on Topological Vector Spaces*, Compact Textbooks in Mathematics,
https://doi.org/10.1007/978-3-030-32945-7_9

(d) If A is complete and $B \subseteq A$ is relatively closed in A, then B is complete. (If $\mathcal{F}$ is a Cauchy filter in B, then $\mathcal{F}$ is a Cauchy filter base in A, which is convergent in A. Since B is closed in A and $B \in \mathcal{F}$, every limit of $\mathcal{F}$ in A belongs to B.)

(e) If E is a topological vector space possessing a countable neighbourhood base of zero $(U_n)_{n \in \mathbb{N}}$, then E is complete if and only if E is sequentially complete. (For the necessity let (x_n) be a Cauchy sequence, i.e., the collection $\{\{x_k ; \ k \geqslant n\}; \ n \in \mathbb{N}\}$ is a Cauchy filter base, and a limit of this filter base is also a limit of the sequence. For the sufficiency let $\mathcal{F}$ be a Cauchy filter. Then there exists a decreasing sequence $(B_n)_n$ in $\mathcal{F}$, $B_n - B_n \subseteq U_n$ $(n \in \mathbb{N})$. For $n \in \mathbb{N}$ choose $x_n \in B_n$. Then (x_n) is a Cauchy sequence, which by hypothesis converges, $x_n \to x$. It is easy to see that then x is a cluster point of $\mathcal{F}$, and therefore $\mathcal{F} \to x$, because $\mathcal{F}$ is a Cauchy filter.)

(f) Let E be a metrisable locally convex space, and let d be a translation invariant metric on E inducing the topology. Then E is complete if and only if the metric space (E, d) is complete (i.e., E is a Fréchet space). This follows immediately from (e) above and the property that $(B_d(0, 1/n))_{n \in \mathbb{N}}$ is a countable neighbourhood base of zero.

(g) Let E, F be topological vector spaces, $u \colon E \to F$ linear and continuous, and let $\mathcal{F}$ be a Cauchy filter in E. Then $\mathrm{fil}(u(\mathcal{F}))$ is a Cauchy filter in F. (If V is a neighbourhood of zero in F, then $u^{-1}(V)$ is a neighbourhood of zero in E. Therefore, there exists $A \in \mathcal{F}$ such that $A - A \subseteq u^{-1}(V)$, and this implies that $u(A) - u(A) \subseteq u(u^{-1}(V)) \subseteq V$.) $\qquad \triangle$

Theorem 9.2
*Let E be a Hausdorff topological vector space. Then there exist a complete Hausdorff topological vector space $\tilde{E}$ such that E is isomorphic to a dense subspace of $\tilde{E}$. The space $\tilde{E}$ is unique up to isomorphism and is called the **completion** of E.*

We will not prove the existence, but rather refer to [Hor66, Chap. 2, § 9, Theorem 1] or [Sch71, Chap. I, § 1.5] for a proof. For locally convex space s we will give a proof later in this chapter. However, we will prove the uniqueness, and for this property we need the following preparations. The first of these is a fundamental fact from topology.

Proposition 9.3 *Let X and Y be topological spaces, Y Hausdorff and regular. Let $X_0 \subseteq X$ be a dense subset, $u_0 \colon X_0 \to Y$ continuous, and suppose that for each $x \in X \setminus X_0$ the limit $u(x) := \lim_{y \to x, \, y \in X_0} u_0(y)$ exists. On X_0 define $u := u_0$. Then u is the unique continuous extension of u_0 to X.*

Recall that **regular** means that every point $y \in Y$ has a neighbourhood base consisting of closed sets. The existence of $\lim_{y \to x, \, y \in X_0} u_0(y)$ means that the image filter base $u_0(\mathcal{U}_x \cap X_0)$ is convergent, where $\mathcal{U}_x$ is the neighbourhood filter of x, and $\mathcal{U}_x \cap X_0 = \{U \cap X_0 ; \ U \in \mathcal{U}_x\}$. The limit is unique because Y is Hausdorff.

Proof of Proposition 9.3

Concerning the uniqueness, assume that u and $\tilde{u}$ are continuous extensions of u_0. Then the set $\{x \in X;\ u(x) = \tilde{u}(x)\}$ is closed (because the diagonal of $Y \times Y$ is closed) and contains X_0, hence is equal to X.

To show the continuity of u, let $x \in X$, and let V be a closed neighbourhood of $u(x)$. By hypothesis, there exists an open neighbourhood U of x such that $u_0(U \cap X_0) \subseteq V$. Then U is a neighbourhood of each of its points z; hence, $u(z) = \lim_{y \to z,\ y \in U \cap X_0} u_0(y) \in \overline{u_0(U \cap X_0)} \subseteq \overline{V} = V$. This shows that $u(U) \subseteq V$ and proves the continuity of u at x. $\qquad\square$

Proposition 9.4 *Let E and F be topological vector spaces, $E_0 \subseteq E$ a dense subspace, F Hausdorff and complete, and let $u_0 \colon E_0 \to F$ be a continuous linear mapping. Then there exists a unique continuous extension $u \colon E \to F$ of u_0, and u is linear.*

Proof

Note that F is regular, because the closed neighbourhoods of zero in F form a neighbourhood base of zero. Let $\mathcal{U}_0$ be the neighbourhood filter of zero in E, and let $x \in E \setminus E_0$. Then

$$\mathcal{F}_x := (x + \mathcal{U}_0) \cap E_0 = \{(x + U) \cap E_0;\ U \in \mathcal{U}_0\}$$

is a filter in E_0 converging to x, hence a Cauchy filter. This implies that $u_0(\mathcal{F}_x)$ is a Cauchy filter base in F, hence convergent. Now Proposition 9.3 yields the existence and uniqueness of the continuous extension u of u_0.

In order to show the linearity of u we let $\lambda \in \mathbb{K}$ and note that the set

$$\{(x, y) \in E \times E;\ u(\lambda x + y) = \lambda u(x) + u(y)\}$$

is a closed subset of $E \times E$ and contains the dense subset $E_0 \times E_0$, hence is equal to $E \times E$. $\qquad\square$

Proof of the uniqueness in Theorem 9.2

Assume that $\tilde{E}$ and $\hat{E}$ are completions, with embeddings $\tilde{\jmath}_0 \colon E \hookrightarrow \tilde{E}$, $\hat{\jmath}_0 \colon E \hookrightarrow \hat{E}$. Interpreting, for the moment, E as a subspace of $\tilde{E}$, we conclude from Proposition 9.4 that $\hat{\jmath}_0$ extends uniquely to $\hat{\jmath} \colon \tilde{E} \to \hat{E}$. Similarly, $\tilde{\jmath}_0$ extends to $\tilde{\jmath} \colon \hat{E} \to \tilde{E}$. As $\tilde{\jmath} \circ \hat{\jmath}$ is continuous, and is the identity on E, it follows that $\tilde{\jmath} \circ \hat{\jmath}$ is the identity on $\tilde{E}$; hence $\hat{\jmath} \colon \tilde{E} \to \hat{E}$ is an isomorphism. $\qquad\square$

The next part of the chapter serves to collect miscellaneous properties concerning completeness.

Proposition 9.5

(a) *Let $(E_\iota)_{\iota \in I}$ be a family of topological vector spaces, and assume that E_ι is (quasi-) complete for all $\iota \in I$. Then $E := \prod_{\iota \in I} E_\iota$ is (quasi-)complete.*

(b) *Let I be a set. Then $\mathbb{K}^I$ is complete.*

(c) *Let E be a vector space. Then $(E^*, \sigma(E^*, E))$ is complete.*

Proof

(a) for 'complete': Let $\mathcal{F}$ be a Cauchy filter in E. Then $\mathrm{pr}_\iota(\mathcal{F})$ is a Cauchy filter base in E_ι, convergent to x_ι ($\iota \in I$). Then $\mathcal{F} \to (x_\iota)_{\iota \in I} \in E$, by Proposition 4.6. The proof for 'quasi-complete' is analogous; observe that, for a bounded set $B \subseteq E$ the images $\mathrm{pr}_\iota(B)$ are bounded ($\iota \in I$).

(b) is a direct consequence of (a).

(c) Recall that E^* is a closed subset of $\mathbb{K}^E$ (Lemma 4.8) and that $\sigma(E^*, E)$ is the restriction of the product topology to E^*. □

Besides being of interest in its own right, the following result serves to prepare the presentation of examples of quasi-complete spaces which are not complete.

Lemma 9.6 *Let E be a barrelled locally convex space. Then $(E', \sigma(E', E))$ is quasi-complete.*

Proof

Let $B \subseteq E'$ be $\sigma(E', E)$-bounded and closed. Then B is equicontinuous (Theorem 6.14), i.e., there exists $U \in \mathcal{U}_0(E)$ such that $B \subseteq U^\circ$. By the Alaoglu–Bourbaki theorem, U° is $\sigma(E', E)$-compact, and therefore complete. (If $\mathcal{F}$ is a Cauchy filter in U°, $\hat{\mathcal{F}}$ a finer ultrafilter, then $\hat{\mathcal{F}}$ is convergent, $\hat{\mathcal{F}} \to x$; therefore x is a cluster point of $\mathcal{F}$, $\mathcal{F} \to x$.) This implies that B is complete. □

Examples 9.7

(a) Let E be a Hausdorff locally convex space, and assume that there exists a linear subspace which is not closed. Then the dual pair $\langle E, E' \rangle$ is separating in E, and passing to the dual pair $\langle E, E^* \rangle$, we note that Corollary 2.10 implies that E' is $\sigma(E^*, E)$-dense in E^*. It is not difficult to show that under the above hypotheses $E' \neq E^*$, and therefore $(E', \sigma(E', E))$ is not complete.

(b) Let E be an infinite-dimensional Banach space. Then $(E', \sigma(E', E))$ is quasi-complete, by Lemma 9.6, but part (a) shows that $(E', \sigma(E', E))$ is not complete. Indeed, it follows from Baire's theorem that countably infinite-dimensional subspaces of E are not closed. △

The following result presents an interesting and surprising interplay concerning completeness in different topologies. It will be important and applied repeatedly in Chapter 14.

Theorem 9.8

Let E be a vector space, let $\sigma \subseteq \tau$ be two linear topologies on E, and assume that τ has a neighbourhood base of zero $\mathcal{U}$ consisting of σ-closed sets.

(a) *Let $\mathcal{F}$ be a τ-Cauchy filter, $x \in E$, $\mathcal{F} \xrightarrow{\sigma} x$. Then $\mathcal{F} \xrightarrow{\tau} x$.*

(b) *Let $A \subseteq E$ be σ-complete. Then A is also τ-complete.*

Proof

(a) Let $U \in \mathcal{U}$. There exists $B \in \mathcal{F}$ such that $B - B \subseteq U$. For $y, z \in B$ one therefore has $y - z \in U$, and as U is σ-closed one obtains $y - x \in U$. This implies that $B \subseteq x + U$, and therefore $\mathcal{F} \xrightarrow{\tau} x$.

(b) This is clear from (a), because every τ-Cauchy filter is a σ-Cauchy filter. $\qquad\square$

The analogous result also holds for the 'sequential setup', with 'closed' replaced by 'sequentially closed', 'Cauchy filter' by 'Cauchy sequence', and 'complete' by 'sequentially complete'.

Example 9.9

Let $1 \leqslant p \leqslant \infty$. On ℓ_p let τ be the norm topology, and let σ be the restriction of the product topology on $\mathbb{K}^{\mathbb{N}}$.

The closed unit ball B_{ℓ_p} is easily seen to be sequentially σ-closed and sequentially σ-complete. Therefore the sequential version of Theorem 9.8 is applicable, and part (b) yields that B_{ℓ_p} (and therefore ℓ_p) is complete.

This (seemingly complicated) proof of the completeness of ℓ_p is nothing but an abstract version of the usual proof of the completeness of ℓ_p. $\qquad\triangle$

The next aim is to prove the following result.

Theorem 9.10

Let E be a quasi-complete locally convex space. Then every $\sigma(E', E)$-bounded subset of E' is $\beta(E', E)$-bounded, i.e., $\mathcal{B}_\beta = \mathcal{B}_\sigma$, in the terminology of the end of Chapter 6.

Before we start with the preparations for the proof we mention a consequence of this result.

Corollary 9.11 *Let E be a quasi-complete quasi-barrelled locally convex space. Then E is barrelled.*

Proof

We will use the terminology of the end of Chapter 6. The fact that E is quasi-barrelled is equivalent to $\mathcal{E} = \mathcal{B}_\beta$ (Theorem 6.8), whereas the quasi-completeness implies that $\mathcal{B}_\beta = \mathcal{B}_\sigma$ (Theorem 9.10). Putting this together we conclude that $\mathcal{E} = \mathcal{B}_\sigma$ which is equivalent to E being barrelled (Theorem 6.14). $\qquad\square$

Let (E, τ) be a locally convex space, and let $B \subseteq E$ be absolutely convex, bounded and closed. Define

$$E_B := \bigcap_{n \in \mathbb{N}} nB = \lim B,$$

with semi-norm p_B. Then $(E_B, p_B) \hookrightarrow (E, \tau)$ is continuous (because B is bounded).

If p_B is a norm and (E_B, p_B) is a Banach space, then B is called a **Banach disc**. Note that p_B is a norm if E is Hausdorff.

Lemma 9.12 *Let E be a locally convex space, and let $B \subseteq E$ be absolutely convex, bounded, closed and sequentially complete.*
(a) *Then (E_B, p_B) is complete. In particular, if p_B is a norm, then B is a Banach disc.*
(b) *Let $D \subseteq E$ be a barrel. Then D absorbs B.*

Proof
(a) follows from the 'sequential version' of Theorem 9.8, applied to E_B, with $\sigma_{E_B} := \tau \cap E_B$, $\tau_{E_B} := \tau_{p_B}$. The conclusion is that the ball $B = \{x \in E_B ; \ p_B(x) \leqslant 1\}$ is p_B-complete.

  (b) (E_B, p_B) is semi-normed and complete, therefore a Baire space (see Appendix B), hence barrelled (Theorem 6.9). The set $D \cap E_B$ is a barrel in (E_B, p_B), therefore a neighbourhood of zero, and therefore absorbs B.   □

Proof of Theorem 9.10
Let $B \subseteq E'$ be $\sigma(E', E)$-bounded. Then B° is a barrel. If $A \subseteq E$ is bounded, then $A^{\circ\circ} = \overline{\mathrm{aco}}\,A$ is closed and bounded, and therefore complete, by hypothesis. Then Lemma 9.12(b) implies that B° absorbs $A^{\circ\circ}$, and therefore $B \subseteq B^{\circ\circ}$ is absorbed by $(A^{\circ\circ})^\circ = A^\circ$. This shows that B is $\beta(E', E)$-bounded.   □

With the following theorem we start the proof of the existence of the completion of a locally convex space; in fact, this theorem is the main ingredient of the proof and also provides a description of the completion.

Theorem 9.13 (Grothendieck)
Let E be a Hausdorff locally convex space. Let $\mathcal{M}$ be a directed covering of E, consisting of bounded, closed, absolutely convex sets. Let

$$F := \{u \in E^* ; \ u|_A \text{ continuous } (A \in \mathcal{M})\}.$$

Then $\mathcal{M}$ can be used to define a polar topology on F in the dual pair $\langle E, F \rangle$, and $(F, \tau_\mathcal{M})$ is a completion of $(E', \tau_\mathcal{M})$.

For the proof we need several preparations.

Lemma 9.14 *Let E be a Hausdorff locally convex space, and let $A \subseteq E$ be absolutely convex and closed. Let $u \in E^*$, $u|_A$ continuous at 0, and let $\varepsilon > 0$. Then there exists $x' \in E'$ such that $|u(x) - \langle x, x' \rangle| \leqslant \varepsilon \ (x \in A)$.*

Proof
It is clearly sufficient to show this for $\varepsilon = 1$.

The continuity of $u|_A$ at 0 implies that there exists an absolutely convex closed neighbourhood of zero $U \subseteq E$ such that $|u(x)| \leqslant 1$ $(x \in A \cap U)$. The polar $U^\bullet$ (taken in the dual pair $\langle E, E^* \rangle$) is a subset of E', $\sigma(E^*, E)$-compact (by the Alaoglu–Bourbaki theorem) and absolutely convex. Therefore Lemma 7.3(b) implies that $A^\bullet + U^\bullet$ is $\sigma(E^*, E)$-closed. Evidently, $A^\bullet + U^\bullet$ is also absolutely convex, and therefore $A^\bullet + U^\bullet = (A^\bullet + U^\bullet)^{\bullet\bullet}$, by the bipolar theorem. Now,

$$u \in (A \cap U)^\bullet = (A^{\bullet\bullet} \cap U^{\bullet\bullet})^\bullet = (A^\bullet \cup U^\bullet)^{\bullet\bullet} \subseteq (A^\bullet + U^\bullet)^{\bullet\bullet} = A^\bullet + U^\bullet.$$

(In the second equality we have used Remark 3.3(c).)

This shows that there exist $w \in A^\bullet$, $x' \in U^\bullet \subseteq E'$ such that $u = w + x'$, and this implies $|u(x) - x'(x)| = |w(x)| \leqslant 1$ for all $x \in A$. $\qquad\square$

Lemma 9.15 *Let X be a topological space, $\mathcal{S} \subseteq \mathcal{P}(X)$. Then the space*

$$C_\mathrm{b}(X, \mathcal{S}) := \{f \colon X \to \mathbb{K} \, ; \; f|_A \text{ bounded and continuous } (A \in \mathcal{S})\},$$

with the semi-norms p_A,

$$p_A(f) := \sup_{x \in A} |f(x)| \quad (f \in C_\mathrm{b}(X, \mathcal{S}), \; A \in \mathcal{S})$$

is complete.

Proof
Without loss of generality we may assume that $\bigcup \mathcal{S} = X$.

For $A \in \mathcal{S}$ the space $C_\mathrm{b}(A)$ (bounded continuous functions with sup-norm) is complete. Let $\mathcal{F}$ be a Cauchy filter in $C_\mathrm{b}(X, \mathcal{S})$. Then for $A \in \mathcal{S}$ the image filter $\mathcal{F}_A$ in $C_\mathrm{b}(A)$ under the mapping $f \mapsto f|_A$ is a Cauchy filter, therefore convergent. This implies that there exists $g \in C_\mathrm{b}(X, \mathcal{S})$ such that $\mathcal{F} \to g$. (Observe that for $A, B \in \mathcal{S}$ with $A \cap B \neq \varnothing$ the limits g_A, g_B of $\mathcal{F}_A, \mathcal{F}_B$ coincide on $A \cap B$. Also, recall Proposition 4.6(b).) $\qquad\square$

Proof of Theorem 9.13
We work in the dual pair $\langle E, F \rangle$.

First we show that $\mathcal{M} \subseteq \mathcal{B}_\sigma(E, F)$ (which makes it clear that $\mathcal{M}$ defines a polar topology on F). Let $A \in \mathcal{M}$, $u \in F$. There exists $U \in \mathcal{U}_0(E)$ such that $|u(x)| \leqslant 1$ $(x \in A \cap U)$. Also, $\lambda A \subseteq U$ for suitable $\lambda \in (0, 1]$ (because A is bounded). For $x \in A$ it follows that $\lambda x \in A \cap U$, $|u(x)| \leqslant \frac{1}{\lambda}$. Therefore A is $\sigma(E, F)$-bounded.

From Lemma 9.14 one concludes that E' is dense in F. (Recall that $\mathcal{M}$ is directed. This implies that $\mathcal{U} := \{\varepsilon B_{q_A} \, ; \; A \in \mathcal{M}, \; \varepsilon > 0\}$ is a neighbourhood base of zero for $\tau_\mathcal{M}$.)

Finally, $(F, \tau_\mathcal{M})$ is complete: $C_\mathrm{b}(E, \mathcal{M})$ is complete, by Lemma 9.15, and $\mathrm{id} \colon C_\mathrm{b}(E, \mathcal{M}) \hookrightarrow \mathbb{K}^E$ is continuous (with the product topology on $\mathbb{K}^E$). Moreover E^* is a closed subspace of $\mathbb{K}^E$ (Lemma 4.8). This shows that $F = C_\mathrm{b}(E, \mathcal{M}) \cap E^*$ is closed in $C_\mathrm{b}(E, \mathcal{M})$, hence complete. $\qquad\square$

Corollary 9.16 **(Grothendieck)** *Let E be a Hausdorff locally convex space, and recall the notation $\mathcal{E} = \{B \subseteq E';\ B\ \text{equicontinuous}\}$. Then*

$$\tilde{E} := \{u \in E'^*;\ u|_B\ \sigma(E', E)\text{-continuous } (B \in \mathcal{E})\},$$

with the polar topology $\tau_{\mathcal{E}}$, is a completion of E. In particular, E is complete if and only if $E = \tilde{E}$.

Proof
This is obtained by applying Theorem 9.13 to $(E', \sigma(E', E))$ and $\mathcal{M} := \{U^\circ;\ U \in \mathcal{U}_0(E)\}$. Then $(E', \sigma(E', E))' = E$, and $\tau_{\mathcal{M}} = \tau_{\mathcal{E}}$ is the original topology on E. $\qquad\qquad\square$

Remark 9.17 If one is just interested in the existence of a completion of a Hausdorff locally convex space E, one can proceed by a reduced method as follows. We only sketch this procedure and refer to [MeVo97, Proposition 22.21] for more details.

With a neighbourhood base of zero $\mathcal{U}$ in E one equips

$$E'^\times := \{u \in E'^*;\ u|_{U^\circ}\ \text{bounded } (U \in \mathcal{U})\}$$

with the semi-norms q_{U°,

$$q_{U^\circ}(u) := \sup\{|\langle u, y\rangle|;\ y \in U^\circ\} \qquad (u \in E'^\times,\ U \in \mathcal{U}).$$

Then $E \subseteq E'^\times$ isomorphically, in the natural way, and $E'^\times$ is complete; hence a completion of E is obtained as $\tilde{E} := \overline{E}^{E'^\times}$. $\qquad\qquad\triangle$

Corollary 9.18 **(Banach)** *Let E be a Banach space, and let $u \in E'^*$ be $\sigma(E', E)$-continuous on $B_{E'}$ (the closed unit ball of E'). Then u belongs to E.*

Proof
By hypothesis, u is $\sigma(E', E)$-continuous on all equicontinuous sets $B \subseteq E'$. Applying Corollary 9.16 and using that E is complete one obtains $u \in E$. $\qquad\qquad\square$

We conclude this chapter with a result on the completeness of dual spaces.

Theorem 9.19
Let E be a bornological locally convex space. Then $(E', \beta(E', E))$ is complete.

We need preparations for the proof.

Lemma 9.20 *Let E be a topological vector space. Then a set $B \subseteq E$ is bounded if and only if, for every sequence $(x_n)_{n \in \mathbb{N}}$ in B and every null sequence $(\lambda_n)_{n \in \mathbb{N}}$ in $\mathbb{K}$, the sequence $(\lambda_n x_n)_{n \in \mathbb{N}}$ is a null sequence.*

Proof

For the necessity, let (x_n) and (λ_n) be as assumed above, and let $U \in \mathcal{U}_0$. There exist $\varepsilon > 0$ such that $\lambda B \subseteq U$ for $|\lambda| \leqslant \varepsilon$, $n_0 \in \mathbb{N}$ such that $|\lambda_n| \leqslant \varepsilon$ $(n \geqslant n_0)$. Then $\lambda_n x_n \in U$ $(n \geqslant n_0)$.

For the sufficiency, assume that B is not bounded. Then there exists $U \in \mathcal{U}_0$ such that $B \not\subseteq nU$ $(n \in \mathbb{N})$. With $x_n \in B \setminus nU$ one obtains $\frac{1}{n}x_n \notin U$ $(n \in \mathbb{N})$, $\frac{1}{n}x_n \not\to 0$. □

Lemma 9.21 *Let E, F be locally convex spaces, $u \colon E \to F$ linear, $B \subseteq E$ bounded and absolutely convex, $u|_B$ continuous at 0. Then $u(B)$ is bounded.*

Proof

Let (x_n) be a sequence in B, (λ_n) a null sequence in $\mathbb{K}$. Then $\lambda_n x_n \in B$ for large n, $\lambda_n x_n \to 0$ by Lemma 9.20, and by hypothesis $\lambda_n u(x_n) = u(\lambda_n x_n) \to 0$ $(n \to \infty)$. Therefore Lemma 9.20 implies that $u(B)$ is bounded. □

Remark 9.22 In Lemma 9.21 (as well as in Lemma 9.14) a linear mapping u was used whose restriction to an absolutely convex set is continuous at 0. It can be shown that the continuity at 0 is equivalent to the continuity on the whole absolutely convex set; cf. [Hor66, Chap. 3, § 11, Lemma 1]. △

Proof of Theorem 9.19

We apply Theorem 9.13 with

$$\mathcal{M} := \big\{ A \subseteq E \, ; \ A \text{ bounded, closed, absolutely convex} \big\};$$

then $\tau_{\mathcal{M}} = \beta(E', E)$.

Let $u \in E^*$, $u|_A$ continuous for all $A \in \mathcal{M}$. By Lemma 9.21, $u(A)$ is bounded for all $A \in \mathcal{M}$, and therefore Proposition 6.18 implies that u is continuous, i.e., $u \in E'$. Now Theorem 9.13 implies that $(E', \beta(E', E))$ is complete. □

Remark 9.23 As metrisable locally convex spaces are bornological, Theorem 9.19 implies that the duals of the following spaces are complete: $C_0^\infty(\Omega)$, $\mathcal{E}(\Omega)$, for open $\Omega \subseteq \mathbb{R}^n$, $\mathcal{S}(\mathbb{R}^n)$, and $C(X)$, for σ-compact Hausdorff locally compact spaces X. △

Notes The material of this chapter, up to Lemma 9.12, is rather standard; Proposition 9.3 is from [Bou07c, Chap. I, § 8.5, Théorème 1, p. I.57]. Theorem 9.8 is an interesting result which can be used to prove completeness of a set if completeness is known in a finer topology; its counterpart for uniform spaces can be found in [Bou07c, Chap. II, § 3.3, Proposition 7 and Corollaire]. Theorem 9.13 and Corollary 9.16 are due to Grothendieck [Gro50]. Following Horváth [Hor66, Chap. 3, § 11, Corollary 4], the author attributes Corollary 9.18 to Banach, although he did not find a direct reference to this result in Banach's publications. However, we will show in Remark 12.3 that it is an immediate consequence of another result of Banach's.

Locally Convex Final Topology, Topology of $\mathcal{D}(\Omega)$

The topic of this chapter is of interest because of its applications to function spaces occurring in partial differential equations. In particular, we describe a neighbourhood base of zero for the space $\mathcal{D}(\Omega)$ of 'test functions'. A further highlight is Köthe's theorem on completeness, Theorem 10.18, which implies that $\mathcal{D}(\Omega)$ is complete.

Let $\Omega \subseteq \mathbb{R}^n$ be an open set. Then we define the **space of test functions**

$$\mathcal{D}(\Omega) := C_c^\infty(\Omega) := \big\{ f \in C^\infty(\Omega) ;\ \operatorname{spt} f \text{ compact} \big\}.$$

For compact $K \subseteq \Omega$ we define

$$\mathcal{D}_K(\Omega) := \big\{ f \in \mathcal{D}(\Omega) ;\ \operatorname{spt} f \subseteq K \big\}\ (= C_0^\infty(\overset{\circ}{K})),$$

a Fréchet space (whose topology we denote by τ_K.) The topology of $\mathcal{D}(\Omega)$ will be defined as the finest locally convex topology for which all the embeddings $\mathcal{D}_K(\Omega) \hookrightarrow \mathcal{D}(\Omega)$ are continuous.

Theorem 10.1

Let E be a vector space, $(X_\iota, \tau_\iota)_{\iota \in I}$ a family of topological spaces, $f_\iota \colon X_\iota \to E$ ($\iota \in I$).

(a) *Then there exists a finest linear (resp. locally convex) topology τ on E, for which all the mappings $f_\iota \colon (X_\iota, \tau_\iota) \to (E, \tau)$ are continuous. τ is called the **linear** (resp., **locally convex**) **final topology**.*

(b) *If (F, σ) is a topological vector space (resp. locally convex space) and $g \colon E \to F$ is a linear mapping, then $g \colon (E, \tau) \to (F, \sigma)$ is continuous if and only if $g \circ f_\iota \colon (X_\iota, \tau_\iota) \to (F, \sigma)$ is continuous for all $\iota \in I$.*

© Springer Nature Switzerland AG 2020
J. Voigt, *A Course on Topological Vector Spaces*, Compact Textbooks in Mathematics,
https://doi.org/10.1007/978-3-030-32945-7_10

Proof

(a) Let Γ be the set of all linear (resp. locally convex) topologies on E, for which all f_ι are continuous; note that $\Gamma \neq \varnothing$ because $\{\varnothing, E\} \in \Gamma$. Then Theorem 1.5 implies that

$$\tau := \text{top} \bigcup \Gamma,$$

the initial topology with respect to the family $(\text{id} \colon E \to (E, \sigma))_{\sigma \in \Gamma}$, is a linear (resp. locally convex) topology on E. Also, $f_\iota \colon (X_\iota, \tau_\iota) \to (E, \tau)$ is continuous ($\iota \in I$), by Theorem 1.2; therefore, $\tau \in \Gamma$.

(b) The necessity of the condition is obvious. To show the sufficiency, let τ' be the initial topology on E with respect to g. Then Theorem 1.2 implies that $f_\iota \colon (X_\iota, \tau_\iota) \to (E, \tau')$ is continuous, for all $\iota \in I$. Therefore $\tau' \in \Gamma$, $\tau \supseteq \tau'$, i.e., $g \colon (E, \tau) \to (F, \sigma)$ is continuous. $\square$

Now, in view of Theorem 10.1 we can define the topology $\tau_\mathcal{D}$ of $\mathcal{D}(\Omega)$ as the locally convex final topology with respect to the family of mappings $\big(\mathcal{D}_K(\Omega) \hookrightarrow \mathcal{D}(\Omega)\big)_{K \subseteq \Omega \text{ compact}}$.

Corollary 10.2 *Let $\Omega \subseteq \mathbb{R}^n$ be open, $u \colon \mathcal{D}(\Omega) \to \mathbb{K}$ linear. Then the following properties are equivalent:*

(i) *u is continuous;*

(ii) *$u|_{\mathcal{D}_K(\Omega)}$ is continuous for all compact $K \subseteq \Omega$;*

(iii) *for each sequence (f_k) in $\mathcal{D}(\Omega)$ such that $\bigcup_{k \in \mathbb{N}} \text{spt } f_k$ is relatively compact, and such that $\partial^\alpha f_k \to 0$ ($k \to \infty$) uniformly on Ω, for all $\alpha \in \mathbb{N}_0^n$, one has $u(f_k) \to 0$ ($k \to \infty$).*

Proof

'(i) $\Leftrightarrow$ (ii)' is a consequence of Theorem 10.1(b).

(ii) $\Leftrightarrow$ (iii). This holds because the convergence stated in (iii) is just the convergence of (f_k) to 0 in $\mathcal{D}_K(\Omega)$, for compact $K \supseteq \bigcup_{k \in \mathbb{N}} \text{spt } f_k$. Since $\mathcal{D}_K(\Omega)$ is a metric space, the property described in (iii) is just the continuity of $u|_{\mathcal{D}_K(\Omega)}$ (at 0). $\square$

Remark 10.3 In the 'theory of distributions without topology' one uses condition (iii) of Corollary 10.2 as the 'continuity condition'. $\mathcal{D}(\Omega)'$ is the space of **distributions** on Ω. The conditions of Corollary 10.2 are further equivalent to

(iv) for all compact $K \subseteq \Omega$ there exist $m \in \mathbb{N}_0$ and $c \geqslant 0$ such that

$$|u(f)| \leqslant c \max \left\{\|\partial^\alpha f\|_\infty; \; |\alpha| \leqslant m\right\} \quad (f \in \mathcal{D}_K(\Omega)).$$

It is obvious that (iv) is equivalent to (ii). $\triangle$

Let E be a vector space, I a directed ordered index set, $(E_\iota)_{\iota \in I}$ a family of subspaces of E, $E_\iota \subseteq E_\kappa$ for $\iota \leqslant \kappa$, $E = \bigcup_{\iota \in I} E_\iota$. For $\iota \in I$ let τ_ι be a locally convex topology on E_ι, and for $\iota \leqslant \kappa$ let $E_\iota \hookrightarrow E_\kappa$ be continuous. Let τ be the locally convex final topology on E with respect to $(E_\iota, \tau_\iota) \hookrightarrow E$ $(\iota \in I)$. Then (E, τ) is called a **locally convex inductive limit**. The inductive limit is called **strict** if $\tau_\iota = \tau_\kappa \cap E_\iota$ for $\iota \leqslant \kappa$.

If (E, τ) is a locally convex inductive limit of a sequence of Banach spaces or of a sequence of Fréchet spaces, then (E, τ) is called an **LB-space** or an **LF-space**, respectively.

Examples 10.4

(a) $\mathcal{D}(\Omega)$ is a strict LF-space: If $(\Omega_k)_{k \in \mathbb{N}}$ is a standard exhaustion of Ω and $K_k := \overline{\Omega_k}$ $(k \in \mathbb{N})$, then $\mathcal{D}(\Omega)$ is the strict locally convex inductive limit of $(\mathcal{D}_{K_k}(\Omega), \tau_{K_k})_{k \in \mathbb{N}}$. (Note that the topology on $\mathcal{D}(\Omega)$ does *not* depend on the choice of the standard exhaustion.)

(b) For $m \in \mathbb{N}_0$, $\Omega \subseteq \mathbb{R}^n$ open, the space

$$\mathcal{D}^m(\Omega) = C_c^m(\Omega) := \left\{ f \in C^m(\Omega); \ \text{spt} f \ \text{compact} \right\}.$$

is a strict LB-space. (Here it is understood that $C^m(\Omega)$ denotes the set of m-times continuously differentiable functions, and that the topology on $C_c^m(\Omega)$ is defined analogously to the topology on $C_c^\infty(\Omega)$.)

(c) Let Ω be a Hausdorff locally compact space. Then $C_c(\Omega)$, with the topology as in (b), for $m = 0$, is the strict inductive limit of $((C_0(\mathring{K}), \| \cdot \|_\infty))_{K \subseteq \Omega \ \text{compact}}$. If Ω is σ-compact, then $C_c(\Omega)$ is a strict LB-space.

(d) Let I be a set. Then $c_c(I) = C_c(I)$ corresponding to (c), with the discrete topology on I, is a strict inductive limit. It is easy to see that $c_c(I)' = \mathbb{K}^I$. (Recall Example 1.7(c).)

(e) Let $E := \mathcal{H}(\{0\})$ be the space of germs of holomorphic functions near 0. Then $\left((\mathcal{H}_b(B(0, \frac{1}{n})), \| \cdot \|_\infty) \right)_{n \in \mathbb{N}}$ is an 'inductive spectrum' for a locally convex inductive limit topology τ on E. In this case (E, τ) is an LB-space, but the inductive limit is not strict. However, here the mappings id: $\mathcal{H}_b(B(0, \frac{1}{n})) \to \mathcal{H}_b(B(0, \frac{1}{n+1}))$ are compact $(n \in \mathbb{N})$, which makes E a 'Silva space'. We refer to [Seb50], [Bar85, Definition 34.1] for more information. $\triangle$

Theorem 10.5

Let E be a vector space, $(E_\iota)_{\iota \in I}$ a family of topological vector spaces, $f_\iota : E_\iota \to E$ linear $(\iota \in I)$, τ the locally convex final topology on E.

(a) *Then*

$$\mathcal{U} := \left\{ U \subseteq E; \ U \ \text{absolutely convex, absorbing}, \ f_\iota^{-1}(U) \in \mathcal{U}_0(E_\iota) \ (\iota \in I) \right\}$$

is a neighbourhood base of zero for τ.

(Continued)

Theorem 10.5 (continued)

(b) *Suppose additionally that $E = \lin\left(\bigcup_{\iota \in I} f_\iota(E_\iota)\right)$, and for each $\iota \in I$ let $\mathcal{U}_\iota$ be a neighbourhood base of zero in E_ι. Then*

$$\check{\mathcal{U}} := \left\{\aco\left(\bigcup_{\iota \in I} f_\iota(U_\iota)\right); \ U_\iota \in \mathcal{U}_\iota \ (\iota \in I)\right\}$$

is a neighbourhood base of zero for τ.

Proof

(a) Let $U \in \mathcal{U}_0(\tau)$ be absolutely convex. Then obviously $U \in \mathcal{U}$.

It remains to show that $\mathcal{U} \subseteq \mathcal{U}_0(\tau)$. Let $U \in \mathcal{U}$. Then (E, p_U) is locally convex, and $f_\iota : E_\iota \to (E, p_U)$ is continuous for all $\iota \in I$. This implies that $\tau \supseteq \tau_{p_U}$, hence $U \in \mathcal{U}_0(\tau)$.

(b) If $U = \aco\left(\bigcup_{\iota \in I} f_\iota(U_\iota)\right) \in \check{\mathcal{U}}$, then $f_\iota^{-1}(U) \supseteq U_\iota \in \mathcal{U}_0(E_\iota)$ $(\iota \in I)$. Also, U is absorbing, because for $x \in E$ there exist a finite set $J \subseteq I$ and $x_\iota \in E_\iota$ $(\iota \in J)$ such that $x = \sum_{\iota \in J} f_\iota(x_\iota)$, which implies that x is absorbed by $\bigcup_{\iota \in J} f_\iota(U_\iota)$. This shows that $U \in \mathcal{U}$.

On the other hand, if $U \in \mathcal{U}$, then for all $\iota \in I$ there exists $U_\iota \in \mathcal{U}_\iota$ such that $U_\iota \subseteq f_\iota^{-1}(U)$, and evidently $U \supseteq \aco\left(\bigcup_{\iota \in I} f_\iota(U_\iota)\right)$. □

Remark 10.6 We refer to [Bou07a, Chap. II, §4, Exerc. 15] for an example where the *linear* final topology of a family of locally convex spaces is not locally convex. However, in [Bou07a, Chap. II, §4, Exerc. 14] the reader is asked to show that the 'strict *linear* inductive limit' of an increasing sequence of locally convex spaces is automatically locally convex. △

Before entering the more detailed discussion of properties of strict inductive limits we first finish the story concerning the duality of the sequence spaces s and t.

Example 10.7

Resuming Example 7.7(a) we recall the "norm" $q_k : \mathbb{K}^{\mathbb{N}} \to [0, \infty]$, $q_k(y) = \sup_{n \in \mathbb{N}} |y_n| n^{-k}$, giving rise to the weighted ℓ_∞-space

$$t_k := \ell_\infty((n^{-k})_{n \in \mathbb{N}}) = \left\{y \in \mathbb{K}^{\mathbb{N}}; \ q_k(y) < \infty\right\}.$$

Then $t = \bigcup_{k \in \mathbb{N}_0} t_k$, the embeddings $(t_j, q_j) \hookrightarrow (t_k, q_k)$, for $0 \leqslant j < k$, are continuous, and we define the corresponding locally convex inductive limit topology τ on t. Applying Theorem 10.5(b), for each $\varepsilon \in (0, \infty)^{\mathbb{N}_0}$ we obtain a τ-neighbourhood of zero

$$U_\varepsilon := \aco\left(\bigcup_{k \in \mathbb{N}_0} B_{q_k}[0, \varepsilon_k]\right),$$

and these U_ε constitute a neighbourhood base of zero for τ when ε runs through $(0, \infty)^{\mathbb{N}}$. Comparing this neighbourhood base of zero with the neighbourhood base of zero for $\beta(t, s)$

obtained in Example 7.7(a), one concludes immediately that $\beta(t, s) \subseteq \tau$. (To make this explicit, the neighbourhoods of zero described in Example 7.7(a) are the $\sigma(t, s)$-closures of the neighbourhoods described above.) Recalling from Example 8.4(a) that s is reflexive, we know that $(t, \beta(t, s))' = s$. Hence, to verify that $\beta(t, s) = \tau$ (i.e., that $\beta(t, s)$ is indeed the locally convex inductive limit topology on t) it now suffices to show that $(t, \tau)' = s$. This is what we will show now, and this will finish our discussion of the rapidly decreasing sequences s.

Defining

$$t_{k,0} := \left\{ y \in \mathbb{K}^{\mathbb{N}} ; \ \lim_{n \to \infty} |y_n| n^{-k} = 0 \right\} \qquad (k \in \mathbb{N}_0),$$

we obtain closed subspaces of t_k satisfying $t_{k,0} \subseteq t_k \subseteq t_{k+1,0}$ $(k \in \mathbb{N}_0)$, and clearly the topology τ on t is also the inductive limit topology defined by the representation $t = \bigcup_{k \in \mathbb{N}_0} t_{k,0}$. Assume that $\eta \in (t, \tau)'$. Then for all $k \in \mathbb{N}$, the functional η belongs to $t'_{k,0}$, and using the duality $c'_0 = \ell_1$ (and suitable isomorphisms between weighted c_0- and ℓ_1-spaces, as in Example 2.19(c)) we conclude that there exists a sequence $(x_n) \in \mathbb{K}^{\mathbb{N}}$ such that $p_k(x) < \infty$. This sequence does not depend on k, and hence $x \in s$.

Having accomplished the aim to find the strong dual of s as the space t with the inductive limit topology, we want to comment on the DF-space properties of t. From Theorem 10.14, proved below, it follows that t is barrelled, a fortiori countably quasi-barrelled. The other property is that t should contain a countable cobase of bounded sets. If $B \subseteq t$ is a bounded set, then there exists $k \in \mathbb{N}_0$ such that $B \subseteq t_k$. We leave this as an exercise to the reader. (This kind of property will be proved in Theorem 10.8(c) for a rather different setting.) Accepting this property, we obtain the sequence $\left(B_{q_k}(0, k) \right)_{k \in \mathbb{N}}$ as a countable cobase of bounded sets. $\triangle$

Now we come to the description of fundamental properties of strict inductive limits.

Theorem 10.8 (Dieudonné–Schwartz)
Let (E, τ) be a strict locally convex inductive limit of an increasing sequence $\left((E_n, \tau_n) \right)_{n \in \mathbb{N}}$ of locally convex subspaces.
(a) *Then $\tau \cap E_n = \tau_n$ $(n \in \mathbb{N})$.*
(b) *If all E_n are Hausdorff, then E is Hausdorff.*
(c) *Assume that E_n is closed in E_{n+1} for all $n \in \mathbb{N}$. Then a set $B \subseteq E$ is τ-bounded if and only if there exists $n \in \mathbb{N}$ such that $B \subseteq E_n$ and B is τ_n-bounded.*

The following lemma is a preparation for the proof.

Lemma 10.9 *Let E be a locally convex space, $F \subseteq E$ a subspace, $V \in \mathcal{U}_0(F)$ absolutely convex. Then there exists $U \in \mathcal{U}_0(E)$ absolutely convex such that $V = U \cap F$. If $x_0 \in E \setminus \overline{F}$, then one can choose U such that $x_0 \notin U$.*

Proof

There exists $\check{U} \in \mathcal{U}_0(E)$ absolutely convex such that $\check{U} \cap F \subseteq V$. Then $U := \mathrm{co}(\check{U} \cup V)$ is absolutely convex (because $\check{U} \cup V$ is balanced), and $U \cap F = V$. (If $x \in \check{U}$, $y \in V$, $0 < t < 1$, $(1 - t)x + ty \in F$, then $x \in F \cap \check{U} \subseteq V$, $(1 - t)x + ty \in V$.)

If $x_0 \notin \overline{F}$, then one can choose $\check{U}$ such that $(x_0 + \check{U}) \cap F = \varnothing$. Then $x_0 \notin U$, because from $x_0 = (1-t)x + ty$ one would obtain $x_0 - (1-t)x \in F \cap (x_0 - (1-t)\check{U}) \subseteq F \cap (x_0 + \check{U})$, which is a contradiction.   □

Proof of Theorem 10.8

(a) '$\subseteq$' holds because $E_n \hookrightarrow E$ is continuous.

'$\supseteq$' Let $n \in \mathbb{N}$, and let $U_n \in \mathcal{U}_0(E_n)$ be absolutely convex. Lemma 10.9 implies that there exists a sequence $(U_k)_{k \geqslant n}$, $U_k \in \mathcal{U}_0(E_k)$ absolutely convex, $U_k = U_{k+1} \cap E_k$ ($k \geqslant n$). Then Theorem 10.5(a) implies that $U := \bigcup_{k \geqslant n} U_k \in \mathcal{U}_0(E)$. Also, one has $U_n = U \cap E_n$.

(b) Let $x \in E$, $x \neq 0$. There exists $n \in \mathbb{N}$ such that $x \in E_n$. As τ_n is Hausdorff, there exists an absolutely convex $U_n \in \mathcal{U}_0(E_n)$ such that $x \notin U_n$. By Lemma 10.9, there exists $U \in \mathcal{U}_0(E)$ such that $U_n = U \cap E_n$, which implies that $x \notin U$.

(c) The sufficiency is clear from the fact that the continuous linear image of a bounded set is bounded.

To prove the necessity by contradiction, assume that $B \subseteq E$ is such that $B \not\subseteq E_n$ ($n \in \mathbb{N}$). Then there exists a sequence (x_k) in B, $n_1 < n_2 < \cdots$, such that $x_k \in E_{n_{k+1}} \setminus E_{n_k}$ ($k \in \mathbb{N}$). Lemma 10.9 implies that there exist a sequence (U_k), $U_k \in \mathcal{U}_0(E_{n_k})$ absolutely convex, $U_k = U_{k+1} \cap E_{n_k}$, $\frac{1}{k} x_k \notin U_{k+1}$. Then $U := \bigcup_{k \in \mathbb{N}} U_k \in \mathcal{U}_0(E)$ by Theorem 10.5(a), but $x_k \notin kU$ ($k \in \mathbb{N}$). This shows that B is not bounded.

As a consequence, if B is bounded, then there exists $n \in \mathbb{N}$ such that $B \subseteq E_n$, and then the relation $\tau_n = \tau \cap E_n$ implies that B is bounded in E_n.   □

Applying Theorem 10.8 we obtain properties of $\mathcal{D}(\Omega)$.

Corollary 10.10 *Let $B \subseteq \mathcal{D}(\Omega)$. Then B is $\tau_\mathcal{D}$-bounded if and only if there exists a compact set $K \subseteq \Omega$ such that $\mathrm{spt}\, f \subseteq K$ for all $f \in B$, and for all $\alpha \in \mathbb{N}_0^n$ one has $\sup_{f \in B} \|\partial^\alpha f\|_\infty < \infty$.*

Corollary 10.11 *Let $\varnothing \neq \Omega \subseteq \mathbb{R}^n$ be open. Then $\mathcal{D}(\Omega)$ is not metrisable.*

Proof

Assume that $\mathcal{D}(\Omega)$ is metrisable. Then there exists a decreasing neighbourhood base of zero $(U_k)_{k \in \mathbb{N}}$. Let (Ω_k) be a standard exhaustion of Ω. For all $k \in \mathbb{N}$ there exists $f_k \in U_k$ such that $\mathrm{spt}\, f_k \cap \Omega_k = \varnothing$. Then $\{f_k; \, k \in \mathbb{N}\}$ is bounded, by construction. This, however, contradicts Corollary 10.10.   □

The following description of a neighbourhood base of zero could already have been given after Theorem 10.5 (but was postponed in favour of the more structural information given previously).

Theorem 10.12 (Schwartz)

Let (Ω_k) be a standard exhaustion of Ω, $\Omega_0 := \emptyset$. For sequences $m = (m_k)_{k\in\mathbb{N}_0}$ in $\mathbb{N}_0$ and $\varepsilon = (\varepsilon_k)_{k\in\mathbb{N}_0}$ in $(0, \infty)$ we define

$$U(m, \varepsilon) := \bigcap_{k\in\mathbb{N}_0} \left\{ f \in \mathcal{D}(\Omega);\ \sup_{x\notin\Omega_k, |\alpha|\leqslant m_k} |\partial^\alpha f(x)| \leqslant \varepsilon_k \right\}.$$

Then

$$\mathcal{U} := \left\{ U(m, \varepsilon);\ m \in \mathbb{N}_0^{\mathbb{N}_0},\ \varepsilon \in (0, \infty)^{\mathbb{N}_0} \right\}$$

is a neighbourhood base of zero for $\tau_{\mathcal{D}}$.

In the proof we will need a partition of unity, whose existence is the issue of the following lemma.

Lemma 10.13 *Let (U_m) be a locally finite open covering of Ω by relatively compact sets. Then there exists a sequence (φ_m) in $\mathcal{D}(\Omega)$ such that $\varphi_m \geqslant 0$, $\operatorname{spt}\varphi_m \subseteq U_m$ ($m \in \mathbb{N}$), $\sum_{m\in\mathbb{N}} \varphi_m = 1$ (where the last sum is a 'locally finite sum').*

Proof

Let (Ω_k) be a standard exhaustion of Ω, $\Omega_0 := \emptyset$.

For $x \in \Omega$ there exist $m \in \mathbb{N}$, $r_x > 0$ such that $B(x, 2r_x) \subseteq U_m$. There exist a sequence (x_j) as well as indices $0 = j_0 < j_1 < j_2 < \cdots$ such that

$$\overline{\Omega_{k+1}} \setminus \Omega_k \subseteq \bigcup_{j=j_k+1}^{j_{k+1}} B(x_j, r_{x_j}) \quad (k \in \mathbb{N}_0),$$

$$\left(\overline{\Omega_{k+1}} \setminus \Omega_k\right) \cap B(x_j, r_{x_j}) \neq \emptyset \quad (j_k + 1 \leqslant j \leqslant j_{k+1},\ k \in \mathbb{N}_0).$$

For $j \in \mathbb{N}$ let $\psi_j \in \mathcal{D}(\Omega)$, $\operatorname{spt}\psi_j = B[x_j, r_{x_j}]$, $\psi_j(x) > 0$ ($x \in B(x_j, r_{x_j})$). For $m \in \mathbb{N}$ we define

$$\tilde{\varphi}_m := \sum_{j\in\mathbb{N}: B(x_j, 2r_{x_j})\subseteq U_m} \psi_j$$

(finite sum, because $U_m \subseteq \Omega_k$ for large k; therefore $B(x_j, 2r_{x_j}) \subseteq U_m$ is only possible for $j \leqslant j_k$). Then $\tilde{\varphi}_m \in \mathcal{D}(\Omega)$, $\tilde{\varphi}_m \geqslant 0$, $\operatorname{spt}\tilde{\varphi}_m \subseteq U_m$ ($m \in \mathbb{N}$), $\tilde{\varphi} := \sum_{m\in\mathbb{N}} \tilde{\varphi}_m \in C^\infty(\Omega)$ (locally finite sum!), $\tilde{\varphi}(x) > 0$ ($x \in \Omega$). With

$$\varphi_m := \frac{\tilde{\varphi}_m}{\tilde{\varphi}} \quad (m \in \mathbb{N})$$

one obtains the assertions. $\qquad\qquad\qquad\qquad\qquad\qquad\qquad\qquad\qquad\qquad\qquad\qquad\square$

Proof of Theorem 10.12

We denote $K_k := \overline{\Omega_k}$ $(k \in \mathbb{N})$.

(i) First we show that $\mathcal{U} \subseteq \mathcal{U}_0(\tau_{\mathcal{D}})$. If $U \in \mathcal{U}$, then U is absolutely convex, absorbing, and $U \cap \mathcal{D}_{K_k}(\Omega)$ is a neighbourhood of zero in $\mathcal{D}_{K_k}(\Omega)$, for all $k \in \mathbb{N}$. Theorem 10.5(a) implies that $U \in \mathcal{U}_0(\tau_{\mathcal{D}})$.

(ii) Now we show that $\mathcal{U}$ is a neighbourhood base of zero. Let $W \in \mathcal{U}_0(\tau_{\mathcal{D}})$ be absolutely convex. Then for $k \in \mathbb{N}_0$ the set $W \cap \mathcal{D}_{K_k}(\Omega)$ is a neighbourhood of zero in $\mathcal{D}_{K_k}(\Omega)$, and therefore there exist $m_k \in \mathbb{N}_0$ and $\delta_k > 0$ such that

$$\left\{ f \in \mathcal{D}_{K_{k+2}}(\Omega);\ p_{m_k}(f) \leqslant \delta_k \right\} \subseteq W$$

(where p_m is defined by $p_m(f) := \sup\{\|\partial^\alpha f\|_\infty;\ |\alpha| \leqslant m\}$). We note that $(\Omega_{k+2} \setminus K_k)_{k \in \mathbb{N}_0}$ is a locally finite open covering of Ω. Using Lemma 10.13 we obtain a subordinate partition of unity (φ_k).

Each $f \in \mathcal{D}(\Omega)$ can be written as

$$f = \sum_{k=0}^{\infty} \frac{1}{2^{k+1}} \left(2^{k+1} \varphi_k f \right)$$

(in fact a finite sum!). If $2^{k+1} \varphi_k f \in W$ for all k, then also $f \in W$, because W is absolutely convex.

For $k \in \mathbb{N}_0$ there exists $c_k > 0$ such that

$$p_{m_k}(2^{k+1} \varphi_k f) \leqslant c_k \sup_{x \in \Omega_{k+2} \setminus K_k, |\alpha| \leqslant m_k} |\partial^\alpha f(x)| \quad (f \in \mathcal{D}(\Omega)).$$

Set $\varepsilon_k := \delta_k / c_k$ $(k \in \mathbb{N}_0)$.

If $f \in U(m, \varepsilon)$, then one obtains

$$p_{m_k}(2^{k+1} \varphi_k f) \leqslant c_k \sup_{x \notin K_k, |\alpha| \leqslant m_k} |\partial^\alpha f(x)| \leqslant c_k \varepsilon_k = \delta_k;$$

therefore, $2^{k+1} \varphi_k f \in W$ $(k \in \mathbb{N}_0)$, hence $f \in W$. This shows that $U(m, \varepsilon) \subseteq W$. $\square$

Theorem 10.14

Let E be a vector space, $(E_\iota)_{\iota \in I}$ a family of locally convex spaces, $f_\iota \colon E_\iota \to E$ linear, and let E carry the locally convex final topology.

Assume that all E_ι are barrelled/quasi-barrelled/bornological. Then E is barrelled/quasi-barrelled/bornological.

Proof

First note that, if $U \subseteq E$ is a barrel, $f_\iota^{-1}(U)$ is a barrel in E_ι ($\iota \in I$).

Also note that, if $U \subseteq E$ is absolutely convex and bornivorous, then $f_\iota^{-1}(U)$ is absolutely convex and bornivorous ($\iota \in I$). (Recall that the image of a bounded set under a continuous linear mapping is bounded.)

In view of these statements and Theorem 10.5(a), the proof of the three assertions is straightforward. $\qquad\square$

Corollary 10.15 $(\mathcal{D}(\Omega)', \beta(\mathcal{D}(\Omega)', \mathcal{D}(\Omega)))$ *is complete.*

Proof

$\mathcal{D}(\Omega)$ is bornological by Theorem 10.14 and Proposition 6.11; therefore, Theorem 9.19 implies that $(\mathcal{D}(\Omega)', \beta(\mathcal{D}(\Omega)', \mathcal{D}(\Omega)))$ is complete. $\qquad\square$

Theorem 10.16
Let E be the strict locally convex inductive limit of a sequence (E_n) of semi-Montel subspaces, and let E_n be closed in E_{n+1} for all $n \in \mathbb{N}$. Then E is a semi-Montel space.

Proof

Let $B \subseteq E$ be bounded and closed. Theorem 10.8 implies that there exists $n \in \mathbb{N}$ such that $B \subseteq E_n$, and B is bounded and closed in E_n. Therefore B is compact in E_n, and also compact in E. (Note that E is Hausdorff because all E_n are Hausdorff by hypothesis.) $\qquad\square$

Corollary 10.17 $\mathcal{D}(\Omega)$ *is a Montel space, in particular reflexive.*

Proof

Let (Ω_k) be a standard exhaustion of Ω, and let $K_k := \overline{\Omega_k}$ ($k \in \mathbb{N}$). According to Example 8.4(b), $\mathcal{D}_{K_k}(\Omega) = C_0^\infty(\overline{K_k})$ is a Montel space. Now the combination of Theorems 10.14 and 10.16 yields the assertion. $\qquad\square$

Finally, we also want to prove that $\mathcal{D}(\Omega)$ is complete. This will be a consequence of the following theorem.

Theorem 10.18 (Köthe)
Let E be the strict locally convex inductive limit of a sequence (E_n) of complete locally convex spaces. Then E is complete.

Proof

Let $\mathcal{F}$ be a Cauchy filter in E. (We note that the fundamental problem in the proof is that $\mathcal{F}$ need not have a 'trace' on any of the E_n's. The 'augmentation' of $\mathcal{F}$ defined in step (i) below is the main idea in the proof.)

(i) The set

$$\{B + V;\ B \in \mathcal{F},\ V \in \mathcal{U}_0(E)\}$$

is a filter base (because $(B+V) \cap (B'+V') \supseteq (B \cap B') + (V \cap V') \neq \varnothing$), and the generated filter $\mathcal{G}$ is a Cauchy filter. Indeed, for $U \in \mathcal{U}_0$ there exist $V \in \mathcal{U}_0$ such that $V + V - V \subseteq U$ and $B \in \mathcal{F}$ with $B - B \subseteq V$, and therefore

$$(B + V) - (B + V) \subseteq V + V - V \subseteq U.$$

Obviously $\mathcal{F} \supseteq \mathcal{G}$. In the following we will show:

There exists $n \in \mathbb{N}$ such that $A \cap E_n \neq \varnothing$ for all $A \in \mathcal{G}$.   (∗)

(This means that $\mathcal{G}$ has a 'trace' on E_n.)

If this is shown, then $\mathcal{G} \cap E_n$ is a Cauchy filter in E_n, therefore convergent; let $x \in E_n$ be a limit. Then x is a cluster point of $\mathcal{G}$, $x \in \bigcap_{A \in \mathcal{G}} \overline{A \cap E_n} \subseteq \bigcap_{A \in \mathcal{G}} \overline{A}$, and because $\mathcal{G}$ is a Cauchy filter one concludes that $\mathcal{G} \to x$. Since $\mathcal{F} \supseteq \mathcal{G}$, one also has that $\mathcal{F} \to x$.

(ii) Now we prove (∗). Assume that (∗) does not hold. Then there exist sequences (B_n) in $\mathcal{F}$, (V_n) in $\mathcal{U}_0$ such that

$$(B_n + V_n) \cap E_n = \varnothing \quad (n \in \mathbb{N});$$ (10.1)

without loss of generality we can assume that V_n is absolutely convex, $V_{n+1} \subseteq V_n$ $(n \in \mathbb{N})$. We define

$$V := \mathrm{co}\left(\bigcup_{n \in \mathbb{N}} (V_n \cap E_n)\right).$$

Then V is absolutely convex, $V \supseteq V_n \cap E_n$ $(n \in \mathbb{N})$, therefore $V \in \mathcal{U}_0$ (by Theorem 10.5(a)), and

$$(B_n + V) \cap E_n = \varnothing \quad (n \in \mathbb{N}).$$ (10.2)

Indeed, $V \subseteq V_n + E_{n-1}$ (because $V_k \cap E_k \subseteq E_{n-1}$ for $k < n$, $V_k \cap E_k \subseteq V_n$ for $k \geqslant n$), and therefore

$$(B_n + V) \cap E_n \subseteq (B_n + V_n + E_{n-1}) \cap E_n = \varnothing,$$

for all $n \in \mathbb{N}$. (An element in the last intersection would be of the form $b_n + v_n + x_{n-1} = x_n$, hence $b_n + v_n = x_n - x_{n-1} \in E_n$, in contradiction to (10.1).)

Now, as $\mathcal{F}$ is a Cauchy filter there exists $B \in \mathcal{F}$ such that $B - B \subseteq V$. Then $B - B_n \cap B \subseteq V$, $B \subseteq B_n \cap B + V \subseteq B_n + V$, and (10.2) implies that $B \cap E_n = \varnothing$ $(n \in \mathbb{N})$; hence $B = \varnothing$, a contradiction. □

Corollary 10.19 $\mathcal{D}(\Omega)$ *is complete.*

Proof
This is an immediate consequence of Theorem 10.18, since $\mathcal{D}(\Omega)$ is a strict LF-space. □

Corollary 10.19 provides – again, see Corollary 10.11 – an argument why $\mathcal{D}(\Omega)$ is not metrisable, because otherwise $\mathcal{D}(\Omega)$ would be a Baire space; see Theorem B.1. However, the representation $\mathcal{D}(\Omega) = \bigcup_k \mathcal{D}_{K_k}(\Omega)$ from Example 10.4(a) shows that $\mathcal{D}(\Omega)$ is a meagre set, and this would be in conflict with Proposition B.2.

Remark 10.20 If E is a strict locally convex inductive limit of a sequence of quasi-complete/sequentially complete locally convex spaces, with E_n closed in E_{n+1} $(n \in \mathbb{N})$, then E is quasi-complete/sequentially complete. This is immediate from Theorem 10.8. △

Notes The space of distributions as the dual of $\mathcal{D}(\Omega)$ was defined by L. Schwartz, and its theory was developed in [Sch66]. Most of the results on locally convex inductive limits presented in this chapter are contained in [DiSc49]. Theorem 10.18 was proved in [DiSc49, Corollaire de Théorème 6] for the case of LF-spaces and generalised as well as provided with a more direct proof by Köthe [Köt50].

Precompact – Compact – Complete

This chapter is a short survey on the technical properties mentioned in the title, for subsets of topological vector spaces and locally convex spaces.

Let E be a topological vector space, $A \subseteq E$. The set A is called **precompact** if for all $U \in \mathcal{U}_0$ there exists a finite set $F \subseteq E$ such that $A \subseteq F + U$.

Remarks 11.1   (a) If A is precompact, then $\overline{A}$ is precompact.

(b) If A is compact, then A is precompact.

(c) Subsets, scalar multiples and finite unions of precompact sets are precompact.

(d) If A is precompact, then A is bounded.

(e) The notion 'precompact' can be defined in the more general framework of uniform spaces.   △

Theorem 11.2

Let E be a topological vector space, $A \subseteq E$. Then the following properties are equivalent:

(i) A is precompact;

(ii)   every filter in A possesses a finer Cauchy filter;

(iii)   every ultrafilter in A is a Cauchy filter.

Proof

The equivalence '(ii) $\Leftrightarrow$ (iii)' is clear, because every filter possesses a finer ultrafilter.

(i) $\Rightarrow$ (iii). Let $\mathcal{F}$ be an ultrafilter in A, and let $U \in \mathcal{U}_0$. Then there exists a finite set $F \subseteq E$ such that $A \subseteq F + U$. Using Remark 4.3(b) one concludes that one of the sets $(x + U) \cap A$ $(x \in F)$ belongs to $\mathcal{F}$. This implies that $\mathcal{F}$ is a Cauchy filter.

© Springer Nature Switzerland AG 2020

J. Voigt, *A Course on Topological Vector Spaces*, Compact Textbooks in Mathematics,

https://doi.org/10.1007/978-3-030-32945-7_11

(ii) $\Rightarrow$ (i). Assume that A is not precompact. Then there exists $U \in \mathcal{U}_0$ such that $A \setminus (F + U) \neq \varnothing$ for all finite sets $F \subseteq E$. Then the collection

$$\{A \setminus (F + U); \ F \subseteq E \text{ finite}\}$$

is a filter base, with a finer Cauchy filter $\mathcal{F}$. There exists a set $B \in \mathcal{F}$ with $B - B \subseteq U$. For $x \in B$ one deduces that $B \subseteq x + U$, $(x + U) \cap A \in \mathcal{F}$. But also $A \setminus (x + U) \in \mathcal{F}$ by construction. Therefore

$$\varnothing = (A \cap (x + U)) \cap (A \setminus (x + U)) \in \mathcal{F},$$

a contradiction. □

Theorem 11.3

Let E be topological vector space, $A \subseteq E$. Then A is compact if and only if A is precompact and complete.

Proof

For the necessity the precompactness is clear. In order to prove the completeness, let $\mathcal{F}$ be a Cauchy filter. Then Proposition 4.5 implies that $\mathcal{F}$ possesses a cluster point $x \in A$. Then $\mathcal{F} \to x$, because $\mathcal{F}$ is a Cauchy filter.

For the sufficiency let $\mathcal{F}$ be an ultrafilter in A. Then $\mathcal{F}$ is a Cauchy filter, by Theorem 11.2, and therefore converges. Now Proposition 4.5 implies that A is compact. □

Theorem 11.4

Let E be a topological vector space, $A \subseteq E$ precompact.
(a) *Then* bal A *is precompact.*
(b) *If E is locally convex, then* aco A *is precompact.*

Proof

(a) Let $U \in \mathcal{U}_0$ be balanced. There exists a finite set $F \subseteq E$ such that $A \subseteq F + U$. Then

$$\text{bal } A = \bigcup_{|\lambda| \leqslant 1} \lambda A \subseteq \bigcup_{|\lambda| \leqslant 1} (\lambda F + \lambda U) \subseteq \text{bal } F + U.$$

The set bal F is compact; therefore, there exists a finite set $B \subseteq E$ such that bal $F \subseteq B + U$, and so

$$\text{bal } A \subseteq \text{bal } F + U \subseteq B + (U + U).$$

(b) Let $U \in \mathcal{U}_0$ be absolutely convex. Then there exists a finite set $F \subseteq E$ such that $A \subseteq F + U$. The set

$$\text{aco } F = \left\{ \sum_{y \in F} \lambda_y y; \; (\lambda_y) \in \mathbb{K}^F, \; \sum_{y \in F} |\lambda_y| \leqslant 1 \right\}$$

is compact, because it is the continuous image of the compact set

$$\left\{ (\lambda_y)_{y \in F} \in \mathbb{K}^F; \; \sum_{y \in F} |\lambda_y| \leqslant 1 \right\}.$$

Therefore there exists a finite set $B \subseteq E$ such that aco $F \subseteq B + U$.

Let $x \in$ aco A. Then $x = \sum_{j=1}^m \mu_j x_j$, with $x_1, \ldots, x_m \in A$, $\sum_{j=1}^m |\mu_j| \leqslant 1$. Then $x_j = y_j + z_j$ with suitable $y_j \in F, z_j \in U$ ($j = 1, \ldots, m$); therefore

$$x = \sum_{j=1}^m \mu_j x_j = \sum_{j=1}^m \mu_j y_j + \sum_{j=1}^m \mu_j z_j \in B + U + U.$$

Hence aco $A \subseteq B + (U + U)$. □

Corollary 11.5 *Let E be a quasi-complete topological vector space, $A \subseteq E$ compact.*
(a) *Then $\overline{\text{bal}}\, A$ is compact.*
(b) *If E is locally convex, then also $\overline{\text{aco}}\, A$ is compact.*

Proof
This is immediate from Remark 11.1(a) and Theorems 11.3 and 11.4. □

Notes Theorems 11.3 and 11.4 are analogous to what is standard in metric spaces. The remaining facts contain useful information and preparations for later results. For the closed convex hull of a compact set in a Banach space, Corollary 11.5 is due to Mazur [Maz30].

The Banach–Dieudonné and Krein–Šmulian Theorems

In this and the following two chapters we discuss some surprising properties concerning the weak topology of Banach spaces. (However, the discussion will not be restricted to Banach spaces!)

For the first result stated below we will give an interesting and motivating application in the subsequent example. The proof of this result and the more genreral Krein–Šmulian theorem requires the consideration of several additional topologies on locally convex spaces.

Theorem 12.1 **(Banach)**
Let E be a Banach space, $F \subseteq E'$ a subspace. Then F is $\sigma(E', E)$-closed if and only if $F \cap B_{E'}$ is $\sigma(E', E)$-closed.

Example 12.2
Let E be a complex Banach space, $\Omega \subseteq \mathbb{C}$ open, $f \colon \Omega \to E$. A 'traditional' result is then Dunford's theorem: f is holomorphic if and only if $x' \circ f$ is holomorphic for all $x' \in E'$ ([Dun38, Theorem 76], [Yos80, Section V.3]). ('Holomorphic' is defined as complex differentiable, and the $\mathbb{C}$-valued theory of functions of one complex variable carries over to E-valued functions, with the result that E-valued holomorphic functions are analytic.) It is relatively standard that the hypothesis in Dunford's theorem can be weakened to the requirement that $x' \circ f$ is holomorphic for all $x' \in F$, where F is an almost norming subspace

© Springer Nature Switzerland AG 2020
J. Voigt, *A Course on Topological Vector Spaces*, Compact Textbooks in Mathematics,
https://doi.org/10.1007/978-3-030-32945-7_12

of E'. Using Theorem 12.1 one can show that even this condition can be replaced by a weaker requirement:

Let $f : \Omega \to E$ be locally bounded, and assume that the set

$$F := \left\{ x' \in E' ; \ x' \circ f \text{ holomorphic} \right\}$$

is separating in E. Then f is holomorphic.

We start the proof by noting that F is a subspace of E', and that the hypothesis implies that F is $\sigma(E', E)$-dense in E'; see Corollary 2.10. Now we show that the closed unit ball of F,

$$B_F = \left\{ x' \in F ; \ \|x'\| \leqslant 1 \right\} = B_{E'} \cap F,$$

is $\sigma(E', E)$-closed. We introduce the mapping $\varphi : E' \to \mathbb{C}^\Omega, x' \mapsto x' \circ f$ and note that φ is continuous with respect to $\sigma(E', E)$ and the product topology on $\mathbb{C}^\Omega$. By Montel's theorem – see Example 8.4(d) –, the set

$$H := \left\{ g : \Omega \to \mathbb{C} \text{ holomorphic} ; \ |g(z)| \leqslant \|f(z)\| \ (z \in \Omega) \right\}$$

is a compact subset of $C(\Omega)$ (provided with the topology of compact convergence); therefore H is closed in $\mathbb{C}^\Omega$. Then the equality

$$B_F = B_{E'} \cap \varphi^{-1}(H)$$

shows that B_F is $\sigma(E', E)$-closed.

Now we conclude from Theorem 12.1 that F is $\sigma(E', E)$-closed, and therefore $F = E'$. Then the assertion follows from Dunford's theorem.

The result quoted above is due to Grosse-Erdmann ([GrE92]). The above elegant proof is a variant of the proof given by Arendt and Nikolski ([ArNi00, Theorem 3.1]); see also [ABHN11, Theorem A.7].   △

For another application of Theorem 12.1, resulting in a generalisation of Pettis' theorem on measurability of Banach space-valued functions we refer to [ABHN11, Corollary 1.3.3].

Remark 12.3 Corollary 9.18 can be derived from Theorem 12.1. Indeed, if $u \in E'^*$ is $\sigma(E', E)$-continuous on $B_{E'}$, then $u^{-1}(0) \cap B_{E'}$ is $\sigma(E', E)$-closed; hence $u^{-1}(0)$ is a $\sigma(E', E)$-closed subspace of E', and u is $\sigma(E', E)$-continuous, i.e., $u \in E$.   △

The proof of Theorem 12.1 will be given at the end of this chapter; the remainder of the chapter is devoted to preparations for the proof of a more general version.

For a locally convex space E we define a topology τ_f on E' by

$$\tau_f := \left\{ A \subseteq E' ; \ A \cap B \in \sigma(E', E) \cap B \text{ for all equicontinuous sets } B \subseteq E' \right\};$$

it is not difficult te check that τ_f is indeed a topology on E'. Expressed differently, we equip the sets $B \in \mathcal{E}$ (the collection of equicontinuous subsets of E') with the trace $\sigma(E', E) \cap B$ of the weak topology and equip E' with the finest topology on E' for which all injections $j_B : B \hookrightarrow E'$ are continuous. If $\mathcal{E}_0$ is a cobase of $\mathcal{E}$, for instance $\mathcal{E}_0 = \{U^\circ ; \ U \in \mathcal{U}\}$ where $\mathcal{U}$ is a neighbourhood base of zero in E, then τ_f is also the finest topology for which all j_B, for $B \in \mathcal{E}_0$, are continuous. (Concerning terminology: A **cobase** of a collection $\mathcal{A}$ of sets is a subcollection $\mathcal{A}'$ of $\mathcal{A}$ such that for all $A \in \mathcal{A}$ there exists $A' \in \mathcal{A}'$ such that $A \subseteq A'$.)

Clearly, a set $A \subseteq E'$ is τ_f-closed if and only if $A \cap B$ is $\sigma(E', E) \cap B$-closed for all B belonging to a cobase $\mathcal{E}_0$ of $\mathcal{E}$.

Proposition 12.4 *Let E be a locally convex space, and let τ_f be the topology on E' defined above. Then $\tau_f \supseteq \tau_c$ (topology of compact convergence, see Chapter 8). The topology τ_f is Hausdorff, translation invariant, and every τ_f-neighbourhood of zero is absorbing and contains a balanced τ_f-neighbourhood of zero.*

Proof

It was shown in Proposition 8.7 that $\tau_c \cap B = \sigma(E', E) \cap B$ for all equicontinuous sets $B \subseteq E'$. As τ_f is the finest topology coinciding with $\sigma(E', E)$ on the equicontinuous sets, it follows that $\tau_f \supseteq \tau_c$.

The topology τ_f is Hausdorff because $\tau_f \supseteq \sigma(E', E)$, and τ_f is translation invariant because the collection of equicontinuous sets and the topology $\sigma(E', E)$ are translation invariant.

Let V be a τ_f-neighbourhood of zero, $x' \in E'$, $B \subseteq E'$ equicontinuous, balanced and containing x'. Then there exists a balanced $\sigma(E', E)$-neighbourhood of zero W such that $W \cap B \subseteq V \cap B$. There exists $\alpha \in (0, 1)$ such that $\lambda x' \in W$ for $|\lambda| \leqslant \alpha$, and therefore

$$\lambda x' \in W \cap B \subseteq V \cap B \subseteq V \quad (|\lambda| \leqslant \alpha).$$

This shows that V is absorbing.

Let U be a τ_f-neighbourhood of zero, and let

$$V := \bigcup \{A \subseteq U ; \ A \text{ balanced}\}$$

be its 'balanced core' (the largest balanced subset of U). Let $B \subseteq E'$ be equicontinuous and balanced. There exists a balanced $W \in \mathcal{U}_0(\sigma(E', E))$ such that $W \cap B \subseteq U \cap B \subseteq U$. Since $W \cap B$ is balanced, one concludes that $W \cap B \subseteq V$, and this implies that $W \cap B \subseteq V \cap B$. This shows that V is a τ_f-neighbourhood of zero. $\qquad\square$

Remark 12.5 Why can one only show 'balanced' in Proposition 12.4(b)? The reason in the proof is that there does not exist an 'absolutely convex core' of sets. In fact, the reason is deeper, because it is known that in general τ_f is not a linear (let alone a locally convex) topology ([Kōm64, § 2]).

The index 'f' in τ_f is historical and probably just stands for 'finest'. $\qquad\triangle$

Theorem 12.6 (Banach–Dieudonné)
Let E be a metrisable locally convex space. Let

$$\mathcal{M}_{\mathrm{ns}} := \big\{\{x_n ; \, n \in \mathbb{N}\} \cup \{0\} ; \, (x_n) \text{ null sequence in } E\big\}.$$

Then $\tau_{\mathrm{f}} = \tau_{\mathrm{c}} = \tau_{\mathrm{ns}} := \tau_{\mathcal{M}_{\mathrm{ns}}}.$

Proof
(i) '$\tau_{\mathrm{f}} \supseteq \tau_{\mathrm{c}} \supseteq \tau_{\mathrm{ns}}$'. The first inclusion is part of Proposition 12.4; the second inclusion holds because every $A \in \mathcal{M}_{\mathrm{ns}}$ is compact.

(ii) '$\tau_{\mathrm{ns}} \supseteq \tau_{\mathrm{f}}$'. Let U be an open τ_{f}-neighbourhood of zero. It suffices to show that there exists $A \in \mathcal{M}_{\mathrm{ns}}$ such that $A^{\circ} \subseteq U$.

Because E is metrisable, there exists a decreasing neighbourhood base of zero $(V_n)_{n \in \mathbb{N}_0}$ in E, $V_0 = E$, and all V_n absolutely convex and closed. In part (iii) of the proof we will show:

For each $n \in \mathbb{N}_0$ there exists a finite set $B_n \subseteq V_n$ such that $A_n^{\circ} \cap V_n^{\circ} \subseteq U$,

$$(*)$$

where $A_n := \bigcup_{k=0}^{n-1} B_k \ (n \in \mathbb{N}_0)$.

Assuming this, we set $A := \big(\bigcup_{k=0}^{\infty} B_k\big) \cup \{0\}$. Then obviously $A \in \mathcal{M}_{\mathrm{ns}}$. Also $A^{\circ} \subseteq A_n^{\circ}$, and therefore $A^{\circ} \cap V_n^{\circ} \subseteq U \ (n \in \mathbb{N})$. From $\bigcap_{n \in \mathbb{N}} V_n = \{0\}$ one obtains $\bigcup_{n \in \mathbb{N}} V_n^{\circ} = E'$, and therefore $A^{\circ} \subseteq U$.

(iii) We prove $(*)$ by induction. For $n = 0$, the assertion holds with $B_0 = \varnothing$. Assume that B_k has been obtained for $k = 0, \ldots, n-1$. We have to find a finite set $B_n \subseteq V_n$ such that $(A_n \cup B_n)^{\circ} \cap V_{n+1}^{\circ} \subseteq U$.

Set $C := V_{n+1}^{\circ} \setminus U$. The polar V_{n+1}° is compact for $\sigma(E', E)$, by the Alaoglu–Bourbaki theorem. Because V_{n+1}° is equicontinuous, the topologies τ_{f} and $\sigma(E', E)$ agree on V_{n+1}°; therefore, V_{n+1}° is also compact for τ_{f}, and as a consequence the closed subset C is compact for τ_{f}. Since $A_n^{\circ} \cap V_n^{\circ} \subseteq U$ and $U \cap C = \varnothing$, we know that $A_n^{\circ} \cap V_n^{\circ} \cap C = \varnothing$. For all $x \in V_n$ the set $\{x\}^{\circ} \cap A_n^{\circ} \cap C$ is a closed subset of C, and

$$\bigcap_{x \in V_n} \big(\{x\}^{\circ} \cap A_n^{\circ} \cap C\big) = \Big(\bigcap_{x \in V_n} \{x\}^{\circ}\Big) \cap A_n^{\circ} \cap C = V_n^{\circ} \cap A_n^{\circ} \cap C = \varnothing.$$

Now the compactness of C implies that the family $\big(\{x\}^{\circ} \cap A_n^{\circ} \cap C\big)_{x \in V_n}$ cannot have the finite intersection property. This means that there exists a finite subset $B_n \subseteq V_n$ such that $\varnothing = B_n^{\circ} \cap A_n^{\circ} \cap C = (A_n \cup B_n)^{\circ} \cap \big(V_{n+1}^{\circ} \setminus U\big)$, hence $(A_n \cup B_n)^{\circ} \cap V_{n+1}^{\circ} \subseteq U$. □

Remark 12.7 The usual way to formulate Theorem 12.6 is to use the topology τ_{pc}, the topology of uniform convergence on the precompact sets of E, instead of τ_{c}. An inspection of the proof of Proposition 8.7 immediately yields that it also shows that $\tau_{\mathrm{pc}} \cap B = \sigma(E', E) \cap B$ for all equicontinuous sets B. This implies that in Theorem 12.6 one also obtains $\tau_{\mathrm{f}} = \tau_{\mathrm{pc}} = \tau_{\mathrm{ns}}$ (which is the traditional assertion in the Banach–Dieudonné theorem). $\triangle$

Let E be a locally convex space,

$$\mathcal{M}_{\mathrm{cc}} := \big\{ A \subseteq E \,;\, A \text{ convex and compact} \big\}.$$

Then $\tau_{\mathrm{cc}} := \tau_{\mathcal{M}_{\mathrm{cc}}}$, the **topology of compact convex convergence**, is a polar topology on E'. Observe that, in view of Lemma 4.9, the set

$$\mathcal{M}'_{\mathrm{cc}} := \big\{ A \subseteq E \,;\, A \text{ absolutely convex and compact} \big\}$$

is a cobase of $\mathcal{M}_{\mathrm{cc}}$, hence $\tau_{\mathcal{M}'_{\mathrm{cc}}} = \tau_{\mathcal{M}_{\mathrm{cc}}}$. Note that $\sigma(E', E) \subseteq \tau_{\mathrm{cc}} \subseteq \mu(E', E)$; therefore $(E', \tau_{\mathrm{cc}})' = b_1(E) (= E$ if E is Hausdorff).

If E is Hausdorff and quasi-complete, then $\tau_{\mathrm{c}} = \tau_{\mathrm{cc}}$ is compatible with the dual pair $\langle E, E' \rangle$.

Theorem 12.8 (Krein–Šmulian)
Let E be a Fréchet space, and let $\mathcal{U}$ be a neighbourhood base of zero in E. Then a convex set $A \subseteq E'$ is $\sigma(E', E)$-closed if and only if $A \cap U^{\circ}$ is $\sigma(E', E)$-closed for every $U \in \mathcal{U}$.

Proof
The necessity is trivial.

For the sufficiency, we recall that A is τ_{f}-closed, which by Theorem 12.6 implies that A is τ_{c}-closed. By the above preliminary remark, $\tau_{\mathrm{c}} = \tau_{\mathrm{cc}}$ is compatible with the dual pair $\langle E, E' \rangle$, and therefore the convex set A is $\sigma(E', E)$-closed as well. $\square$

Proof of Theorem 12.1
This follows immediately from Theorem 12.8. $\square$

Notes Theorem 12.1 is contained in [Ban32, Chap. VIII, § 3, Lemme 3]. In order to understand this it should be mentioned that the subspaces of E' whose intersection with the closed unit ball is $\sigma(E', E)$-closed occur in [Ban32] as 'transfiniment fermé', whereas $\sigma(E', E)$-closed subspaces are 'régulièrement fermé'. A translation into more modern terminology was given by Bourbaki [Bou38], and a new proof was given by Dieudonné [Die42, Théorème 23]. (Interestingly enough, the proof by contraposition in [Ban32, Chap. VIII, § 3, Lemme 2] seems to have persisted in the literature, where usually in step (iii) of the proof of Theorem 12.6, the existence of a finite set B_n is shown by contraposition.) The new methods introduced by Dieudonné then served to extend Theorem 12.8 – proved in [KrŠm40, Theorem 5] only for the case of Banach spaces – to more general settings. For this and a variety of related results obtained by these methods we refer to Köthe [Köt66, § 21.10] and Schaefer [Sch71, Chap. IV, § 6.4].

The Eberlein–Šmulian and Eberlein–Grothendieck Theorems

The well-known Eberlein–Šmulian theorem states the equivalence of several versions of weak compactness for subsets of a Banach space. The proof will be a consequence of properties of subsets $H \subseteq C(X)$ with respect to the product topology on $\mathbb{K}^X$, where X is a suitable topological space. These considerations will also yield results for more general locally convex spaces.

To motivate the considerations presented in this chapter, we start by stating an important result which will be proved, in fact in a more general version.

Concerning the terminology in the following theorem: If X is a topological space, a subset A is called **conditionally countably compact** if every sequence in A has a cluster point in X. The set A is called **conditionally sequentially compact** if every sequence in A possesses a convergent subsequence with limit in X. This is in contrast to the definition of relative (countable or sequential) compactness, which means that there exists a superset of A with the corresponding property.

Theorem 13.1 (Eberlein–Šmulian)
Let E be a Banach space, $A \subseteq E$. Then the following properties are equivalent:
(i) *A is weakly relatively compact;*
(ii) *A is weakly conditionally countably compact;*
(iii) *A is weakly conditionally sequentially compact.*

If one of these properties is satisfied, then for every $x \in \overline{A}^{\sigma(E,E')}$ there exists a sequence in A converging weakly to x.

Remarks 13.2 (a) Recall that for separable Banach spaces, part of this theorem was already proved in Proposition 5.11.

© Springer Nature Switzerland AG 2020
J. Voigt, *A Course on Topological Vector Spaces*, Compact Textbooks in Mathematics,
https://doi.org/10.1007/978-3-030-32945-7_13

(b) As a consequence of Theorem 13.1 one deduces that, for subsets A of E, weak compactness, weak countable compactness and weak sequential compactness are equivalent properties. (Recall that A is countably compact if every sequence in A has a cluster point in A, and that A is sequentially compact if every sequence has a convergent subsequence with limit belonging to A.)

Indeed, in view of Theorem 13.1 it suffices to show that weak countable compactness implies weak closedness. So, assume that A is weakly countably compact, and let $x \in \overline{A}^{\sigma(E,E')}$. Then by the final assertion of the theorem, there exists a sequence (x_n) in A converging weakly to x. By hypothesis, $x \in A$, and this implies that A is closed.

(c) We note that the assertion stated in Theorem 13.1 also holds in (non-complete) normed spaces. It is not difficult to deduce this version from Theorem 13.1 itself. We mention this fact in order to prepare the reader for later versions of the Eberlein–Šmulian theorem, where completeness will not be required.   △

For the further development we present the following important and surprising result on various notions of compactness in function spaces.

Theorem 13.3 (Eberlein–Grothendieck)
Let X be a topological space having a dense σ-compact subset, and let τ_s be the product topology on $\mathbb{K}^X$. Let $H \subseteq C(X)$ be a subset which is conditionally countably compact with respect to $\tau_s \cap C(X)$ (i.e., every sequence in H has a cluster point in $C(X)$).
Then the set $\overline{H} \;(= \overline{H}^{\tau_s})$ is compact and contained in $C(X)$. Moreover, for every separable set $F \subseteq \overline{H}$ the topology $\tau_s \cap F$ is metrisable, and each $f \in \overline{H}$ is the limit of a sequence in H.

Remarks 13.4 (a) We point out that under the hypotheses of the previous theorem the set H will also be conditionally sequentially compact: Let (f_n) be a sequence in H, and let $f \in \overline{\{f_n\,;\ n \in \mathbb{N}\}}$. If f occurs infinitely often in the sequence (f_n), then there exists a subsequence with all the terms equal to f. Otherwise we may without restriction assume that f does not occur in the sequence (f_n). Then, by the last statement of the theorem, there exists a sequence (g_n) in $\{f_n\,;\ n \in \mathbb{N}\}$ converging to f. Since (g_n) cannot stay in any set $\{f_k\,;\ 1 \leqslant k \leqslant n\}$ for some $n \in \mathbb{N}$, it follows that by 'thinning out' (g_n) one can get a subsequence of (f_n) converging to f. (Note that – for puristic reasons – we did not use the fact that every countable set has a metrisable closure; this would have made the argument somewhat easier.)

(b) We mention that the hypothesis of Theorem 13.3 is satisfied, for instance, if X is separable.   △

Seeing the result of Theorem 13.3 for the first time, each mathematician specialising in analysis will (and should) be surprised. After all, function spaces are one of the main objects in analysis, and so he might have the impression that he 'is somewhat acquainted' with function spaces. However, mostly one doesn't consider the product topology on such spaces; it will turn out that the result – whose proof is somewhat technical and lengthy, even for the special case where X is compact – has very interesting and important consequences. Also, the Eberlein–Šmulian theorem mostly is a theorem quoted and used every now and then in courses of functional anlysis, but – the author thinks – rarely proved completely. Learning that it is just a consequence of sophisticated properties of function spaces should be a good experience.

We need a preparation for part of the proof.

Lemma 13.5 *Let Y, Z be sets, $\langle \cdot, \cdot \rangle \colon Y \times Z \to \mathbb{K}$. Let ρ be the initial topology on Y with respect to the family $(\langle \cdot, z \rangle)_{z \in Z}$, and let σ be the initial topology on Z with respect to $(\langle y, \cdot \rangle)_{y \in Y}$. Assume that (Y, ρ) is compact, and let $Z_0 \subseteq Z$ be dense in (Z, σ).*

Then the initial topology on Y with respect to $(\langle \cdot, z \rangle)_{z \in Z_0}$ is equal to ρ. If Z_0 is countable, then ρ is associated with a semi-metric, and (Y, ρ) is separable.

Proof

Let $\Phi \colon Y \to \mathbb{K}^Z$ be defined by $\Phi(y) := (\langle y, z \rangle)_{z \in Z}$. Then Φ is continuous; in fact Theorem 1.2 implies that ρ is the initial topology with respect to Φ. Therefore the hypothesis on (Y, ρ) implies that $A := \Phi(Y)$ is a compact subset of $\mathbb{K}^Z$. Let $\mathrm{pr}_{Z_0} \colon \mathbb{K}^Z \to \mathbb{K}^{Z_0}$ be the canonical projection. Then the restriction of pr_{Z_0} to A is injective. Indeed, if $x, y \in Y$ are such that $\mathrm{pr}_{Z_0}(\Phi(x)) = \mathrm{pr}_{Z_0}(\Phi(y))$, i.e., $\langle x, z \rangle = \langle y, z \rangle$ ($z \in Z_0$), then the continuity of the functions $\langle x, \cdot \rangle$, $\langle y, \cdot \rangle$ on (Z, σ) implies that $\langle x, z \rangle = \langle y, z \rangle$ ($z \in Z$), i.e., $\Phi(x) = \Phi(y)$. This implies that the restriction of pr_{Z_0} to A is a homeomorphism; cf. Lemma 4.12. As a consequence, the topology ρ is the initial topology with respect to $\mathrm{pr}_{Z_0} \circ \Phi$, or equivalently, by Theorem 1.2, the initial topology with respect to $(\langle \cdot, z \rangle)_{z \in Z_0}$.

If Z_0 is countable, then Lemma 2.18 implies that ρ is semi-metrisable. A standard argument shows that then (Y, ρ) is separable. □

Proof of Theorem 13.3

(i) For all $t \in X$ the set $\{f(t);\ f \in H\}$ is bounded. Therefore the compactness of $\overline{H}$ is a consequence of Tikhonov's theorem.

The proof of the other properties will be given in three parts, where in the first part we prove the theorem for the case that X is compact.

(ii) Let $F_0 \subseteq H$ be countable, $F := \overline{F_0}^{\tau_s \cap C(X)}$. Then the initial topology τ_F on X with respect to F is coarser than the original topology of X, and therefore (X, τ_F) is compact. Applying Lemma 13.5 with $Y := X$, $Z := F$, $Z_0 := F_0$, $\langle t, f \rangle := f(t)$ $((t, f) \in X \times F)$, we conclude that τ_F is equal to the initial topology τ_{F_0} with respect to F_0, and that (X, τ_F) is semi-metrisable and separable.

(iii) Now we show that $\overline{H} \subseteq C(X)$. Assume that there exists $g \in \overline{H}$ which is not continuous. Then there exist $t_0 \in X$, $\varepsilon > 0$, $M \subseteq X$ with $t_0 \in \overline{M}$ such that

$$|g(t) - g(t_0)| \geqslant \varepsilon \quad (t \in M). \tag{13.1}$$

Then one can construct sequences $(f_n)_{n \in \mathbb{N}}$ in H, $(t_n)_{n \in \mathbb{N}}$ in M such that

$$|f_n(t_k) - g(t_k)| \leqslant 1/n \quad (0 \leqslant k < n), \tag{13.2}$$

$$|f_k(t_n) - f_k(t_0)| \leqslant 1/n \quad (1 \leqslant k \leqslant n). \tag{13.3}$$

(Indeed, if $n \in \mathbb{N}$ and $t_1, \dots, t_{n-1}, f_1, \dots, f_{n-1}$ are chosen, then there exists $f_n \in H$ satisfying (13.2), because g belongs to the closure of H. Then $t_n \in M$ satisfying (13.3) can be chosen, because $f_1, \dots, f_n$ are continuous.) By hypothesis, the sequence (f_n) has a cluster point $f \in C(X)$. Then $f(t_k) = g(t_k)$ $(k \in \mathbb{N}_0)$ because of the inequalities (13.2). From (13.3) we deduce that $f_k(t_n) \to f_k(t_0)$ $(n \to \infty)$ for all $k \in \mathbb{N}$, and therefore $t_n \to t_0$ $(n \to \infty)$ in the initial topology τ_{F_0} of X with respect to $F_0 := \{f_k\,;\; k \in \mathbb{N}\}$. From (ii) we know that f is continuous with respect to τ_{F_0}, and therefore $g(t_n) = f(t_n) \to f(t_0) = g(t_0)$ $(n \to \infty)$. This convergence contradicts (13.1).

(iv) Let F_0, F and τ_F be as in (ii). Note that from (i) and (iii) we know that $F = \overline{F_0}$ is a τ_s-compact subset of $C(X)$. In this step we show that (F, τ_s) is metrisable. This implies that F, and thus $\overline{H}$, is sequentially compact.

From (ii) we know that (X, τ_F) is separable. Applying Lemma 13.5 once more, this time with (exchanged roles) $Y := F$, $Z := X$, $\langle f, t \rangle := f(t)$ $((f, t) \in F \times X)$, we conclude that F is a compact (semi-)metrisable space.

(v) Finally we show that, given $f \in \overline{H}$, there exists a sequence (f_n) having f as a cluster point. Then, applying the previous step, one also obtains a convergent subsequence. Let $k \in \mathbb{N}$. For $g \in H$ we define

$$U_g := \big\{(t_1, \dots, t_k) \in X^k\,;\; |g(t_j) - f(t_j)| < 1/k \ (1 \leqslant j \leqslant k)\big\}.$$

Then $(U_g)_{g \in H}$ is an open covering of the compact set X^k, and therefore there exists a finite subcovering $(U_g)_{g \in H_k}$, with a finite set $H_k \subseteq H$. Arranging the countable set $\bigcup_{k \in \mathbb{N}} H_k$ as a sequence (f_n) one easily deduces that this sequence has f as a cluster point.

Now we start the second part of the proof, where we assume that X is σ-compact. Let (X_n) be a sequence of compact sets whose union is X; without restriction we may assume that the sequence (X_n) is increasing.

(vi) In the first step we show that $\overline{H} \subseteq C(X)$ and that each element of $\overline{H}$ is the limit of a sequence in H.

Let $f \in \overline{H}$. Then, for $n \in \mathbb{N}$, the function $f|_{X_n}$ belongs to the closure of $\big\{g|_{X_n}\,;\; g \in H\big\}$ with respect to the product topology of $\mathbb{K}^{X_n}$. The first part of the proof implies that there is a sequence $\big(f_j^n\big)_j$ in H such that $\big(f_j^n|_{X_n}\big)_j$ converges to $f|_{X_n}$ in the product topology of $\mathbb{K}^{X_n}$. Arranging the sequences $\big(f_j^n\big)_j$ $(n \in \mathbb{N})$ into a single sequence (by the usual Cantor counting procedure) we obtain a sequence (f_k) with the property that $\big(f_k|_{X_n}\big)_k$ has the function $f|_{X_n}$ as a cluster point, for all $n \in \mathbb{N}$. It further follows from the first part of the proof that the restriction of the product topology on $\mathbb{K}^{X_n}$ to the (countable) set $\big\{f_k|_{X_n}\,;\; k \in \mathbb{N}\big\} \cup \{f|_{X_n}\}$ is metrisable. Since the product topology on $\mathbb{K}^X$ is the initial topology with respect to the canonical projections $\mathrm{pr}_{X_n} : \mathbb{K}^X \to \mathbb{K}^{X_n}$ ($n \in \mathbb{N}$), we conclude, using Lemma 2.18, that the restriction of the product topology on $\mathbb{K}^X$ to $\{f_k \,;\, k \in \mathbb{N}\} \cup \{f\}$ is metrisable. This, finally, shows that there exists a subsequence $(f_{k_j})_j$ such that $f_{k_j} \to f$ ($j \to \infty$) in the product topology of $\mathbb{K}^X$.

Now from the hypothesis that each sequence in H has a cluster point in $C(X)$ we conclude that the unique cluster point f of $(f_{k_j})_j$ does in fact belong to $C(X)$.

(vii) Let $F_0 \subseteq H$ be countable, $F := \overline{F_0}^{\tau_s}$. Then part one of the proof implies that the product topology of $F|_{X_n}$ is metrisable, for all $n \in \mathbb{N}$, and as the product toplogy on F is the initial topology with respect to the canonical projections $\mathrm{pr}_{X_n} : \mathbb{K}^X \to \mathbb{K}^{X_n}$ ($n \in \mathbb{N}$) it follows that the product topology on F is metrisable. This also shows that F and hence $\overline{H}$ is sequentially compact.

In this third part of the proof we treat the general case. We will use the notation $\check{X} := \bigcup_{n \in \mathbb{N}} X_n$, where (X_n) is a sequence of compact subsets whose union is dense in X.

(viii) First we show that $\overline{H} \subseteq C(X)$ and that for each $f \in \overline{H}$ there exist a sequence converging to f.

Let $f \in \overline{H}$. From the previous part of the proof we know that $f|_{\check{X}}$ is continuous, and that there exists a sequence (f_k) in H such that $f_k|_{\check{X}} \to f|_{\check{X}}$ ($k \to \infty$) pointwise. From the hypothesis we know that (f_k) has a cluster point $g \in C(X)$. This implies that $g|_{\check{X}} = f|_{\check{X}}$, and because $\check{X}$ is dense, the continuous cluster point is unique. Then it is standard to deduce that $f_k \to g$ pointwise on X. (Indeed, assuming that there exists $t \in X$ such that $(f_k(t))$ does not converge to $g(t)$, one would obtain a subsequence (f_{k_j}) with $(f_{k_j}(t))$ converging to some value $\neq g(t)$. However, this subsequence could no longer have a continuous cluster point; a contradiction.) Choosing $t \in X$ arbitrarily, we now invoke part two of the proof once more to conclude that $f|_{\check{X} \cup \{t\}}$ is continuous, hence $f(t) = g(t)$. This implies that $f = g \in C(X)$.

(ix) Let $\mathrm{pr}_{\check{X}} : \mathbb{K}^X \to \mathbb{K}^{\check{X}}$ be the canonical projection, and denote the initial topology on $\overline{H}$ with respect to $\mathrm{pr}_{\check{X}}$ by $\check{\tau}_s$. Then id: $(\overline{H}, \tau_s) \to (\overline{H}, \check{\tau}_s)$ is continuous, and as $\overline{H}$ is compact and $\check{\tau}_s$ is Hausdorff, we obtain $\tau_s \cap \overline{H} = \check{\tau}_s$ (by Lemma 4.12). This makes it clear that, for countable sets $F_0 \subseteq H$, the topology τ_s on $F := \overline{F_0}$ is metrisable, because it holds for the topology $\check{\tau}_s$, by part two of the proof. This also implies that H is conditionally sequentially compact. $\qquad\qquad\square$

As a first application we treat a case where there is no restriction on the quality of the space, but for the subset one does not conclude conditional sequential compactness. For this result, Theorem 13.3 is only used for the case that X is compact.

Corollary 13.6 (Eberlein) *Let E be a Hausdorff locally convex space, and let $A \subseteq E$ be such that $\overline{\mathrm{co}}\, A$ is complete. Then A is weakly relatively compact if and only if A is weakly conditionally countably compact.*

Proof

The necessity is clear.

For the sufficiency, let $\tilde{E}$ be the completion of E. Then $\overline{\text{co}}\, A$ is closed in $\tilde{E}$, hence also $\sigma(\tilde{E}, E')$-closed in $\tilde{E}$; note that $\tilde{E}' = E'$. Therefore it is sufficient to show that $\overline{A}^{\sigma(\tilde{E},E')}$ is weakly compact in $\tilde{E}$; hence, without less of generality we may assume that E is complete.

The set $\overline{A}^{\sigma(E'^*,E')}$ is compact in $(E'^*, \sigma(E'^*, E'))$, by Tikhonov's theorem and the property that E'^* is a closed subset of $\mathbb{K}^{E'}$ (Lemma 4.8). Therefore, we have to show that $\overline{A}^{\sigma(E'^*,E')} \subseteq E$.

Let $X \subseteq E'$ be equicontinuous and $\sigma(E', E)$-closed; then X is $\sigma(E', E)$-compact, by the Alaoglu–Bourbaki theorem. Then we apply Theorem 13.3 to the restrictions of the elements of $A \subseteq E'^* \subseteq \mathbb{K}^{E'}$ to X, to obtain

$$\overline{A}^{\sigma(E'^*,E')}\big|_X = \overline{A|_X}^{\mathbb{K}^X} \subseteq C(X).$$

The equality in this chain holds because the set $\overline{A}^{\sigma(E'^*,E')}$ is compact and the mapping $E'^* \ni x \mapsto x|_X \in \mathbb{K}^X$ is continuous. The inclusion is the application of Theorem 13.3; notice that the hypothesis of conditional countable compactness of A implies that $A|_X = \{x|_X \,;\, x \in A\}$ is a conditionally countably compact subset of $C(X)$.

Therefore it follows that for each element $x \in \overline{A}^{\sigma(E'^*,E')}$ the restriction $x|_X$ is continuous for all closed equicontinuous sets $X \subseteq E'$. This implies that $x \in E$, by Corollary 9.16.   □

Now we show a generalisation of the original Eberlein–Šmulian theorem. The following version applies, in particular, to the weak topology in Fréchet spaces.

Theorem 13.7 (S. Dierolf)

Let (E, τ) be a Hausdorff locally convex space, and assume that there exists a metrisable locally convex topology $\rho \subseteq \mu(E, E')$ on E. Then, for a set $A \subseteq E$, the following properties are equivalent:

  (i)   *A is relatively compact;*
 (ii)   *A is conditionally countably compact;*
(iii)   *A is conditionally sequentially compact.*

If one of these properties is satisfied, then $\overline{A} = \overline{A}^{\sigma}$, where $\sigma := \sigma(E, E')$, $\tau \cap \overline{A} = \sigma \cap \overline{A}$, every separable subset of A is metrisable, and every $x \in \overline{A}$ is the limit of a sequence in A.

The proof will be given in two parts, where in the first part we suppose that τ is the weak topology.

Proof of Theorem 13.7, for $\tau = \sigma \; (= \sigma(E, E'))$

The implications '(i) $\Rightarrow$ (ii)' and '(iii) $\Rightarrow$ (ii)' are clear.

(ii) $\Rightarrow$ (i), (iii). The space (E, ρ) possesses a countable neighbourhood base of zero (U_n). The sets $X_n := U_n^\circ$ ($n \in \mathbb{N}$) are $\sigma(E', E)$-compact, by the Alaoglu–Bourbaki theorem and because $\rho \subseteq \mu(E, E')$, and $F := \bigcup_{n\in\mathbb{N}} X_n = \bigcup_{n\in\mathbb{N}} U_n^\circ = (E, \rho)'$ is a subspace of E'. Since ρ is Hausdorff, the dual pair $\langle E, F \rangle$ is separating in E, and this implies that F is $\sigma(E', E)$-dense in E' (by Corollary 2.10).

Then Theorem 13.3, together with Remark 13.4(a), implies that A, as a subset of $C(E', \sigma(E', E))$, is weakly relatively compact and weakly conditionally sequentially compact. Recalling that the closure of A in $\mathbb{K}^{E'}$ is a subset of E'^*, by Lemma 4.8, we obtain the desired assertions.

The last assertions of the theorem are properties stated in Theorem 13.3. □

For the proof of the general case we need two preparations.

Lemma 13.8 *Let X be a topological space, (x_n) a sequence in X, $\{x_n ; \, n \in \mathbb{N}\}$ countably compact, and such that (x_n) has only one cluster point $x \in X$. Then $x_n \to x$ $(n \to \infty)$.*

Proof

There exists a cluster point x of (x_n). If (x_n) does not converge to x, then there exist an open neighbourhood U of x and a subsequence (x_{n_j}) in $X \setminus U$. This subsequence has a cluster point $y \in X \setminus U$, and $y \neq x$ is also a cluster point of (x_n), in contradiction to the hypothesis. □

Proposition 13.9 *Let E be a topological vector space, and let $A \subseteq E$ be a conditionally countably compact subset. Then A is precompact.*

Proof

The proof proceeds by contraposition. Assume that A is not precompact. Then there exist $U \in \mathcal{U}_0$ and a sequence (x_n) in A such that $x_{n+1} \notin \bigcup_{j=1}^n (x_j + U)$ for all $n \in \mathbb{N}$. There exists $V \in \mathcal{U}_0$ such that $V - V \subseteq U$. It is easy to see that then for all $x \in E$ the set $x + V$ contains at most one point of the sequence (x_n). This shows that (x_n) has no cluster point. □

Proof of Theorem 13.7, general case

As above, the implications '(i) $\Rightarrow$ (ii)' and '(iii) $\Rightarrow$ (ii)' are clear.

Now we show the implication '(ii) $\Rightarrow$ (iii)'. Let (x_n) be a sequence in A. The hypothesis implies that A is weakly conditionally countably compact. Then the first part of the proof implies that A is weakly conditionally sequentially compact. Hence there exists a weakly convergent subsequence (x_{n_j}) of (x_n); let $x := \sigma\text{-}\lim x_{n_j}$. Then every τ-cluster point of (x_{n_j}) is also a σ-cluster point, hence is equal to x. Applying Lemma 13.8 we conclude that $x = \tau\text{-}\lim x_{n_j}$.

Next we show the implication '(ii) $\Rightarrow$ (i)'. As before, A is weakly conditionally countably compact, and from the first part of the proof we know that $\overline{A}^{\sigma}$ is weakly compact, hence weakly complete. Now Theorem 9.8 implies that $\overline{A}^{\sigma}$ is τ-complete. Proposition 13.9 implies that A, and then also $\overline{A}$, is precompact, and therefore $\overline{A}$ is compact, by Theorem 11.3.

Having shown these two implications, we use that the compactness of $\overline{A}$ implies the weak compactness, $\overline{A}^{\sigma} = \overline{A}$, and $\sigma \cap \overline{A} = \tau \cap \overline{A}$, by Lemma 4.12. This implies that the last assertions of the theorem carry over from σ to τ. □

Remarks 13.10 (a) Remark 13.2(b) applies analogously to Theorem 13.7.

(b) The topology ρ in Theorem 13.7 is compatible with the dual pair $\langle E, E' \rangle$, i.e., $\sigma(E, E') \subseteq \rho$, if and only if $\rho = \mu(E, E')$; see Remark 6.17. △

Proof of Theorem 13.1

This is a special case of Theorem 13.7. □

Remark 13.11 We mention that for the proof of Theorem 13.1 it would be sufficient to have Theorem 13.3 for compact X. Indeed, the Banach space E is isometrically isomorphic to the closed subspace $E'^* \cap C(B_{E'}, \sigma(E', E))$ of E'^*, by Corollary 9.18, and the weak topology on E in the isomorphic image is the restriction of the product topology on $\mathbb{K}^{B_{E'}}$ to this subspace. Taking into account that E'^* is closed in the product topology (Lemma 4.8), one obtains the assertions of Theorem 13.1 from Theorem 13.3. △

Remark 13.12 The Eberlein–Šmulian theorem has given rise to investigations concerning the question in which spaces the notions of compactness, countable compactness, and sequential compactness are equivalent. Above, we have presented some results that go beyond the classical Eberlein–Šmulian theorem. A much deeper and more thorough investigation is carried out in [Flo80]. In particular, we mention that these investigations have led to the notion of 'angelic spaces'. The assertions of Theorem 13.1 and Theorem 13.7 imply that the weak topology on the space E is 'strictly angelic', in the terminology of [Gov80]. △

For completeness we mention that Theorem 13.3 also yields part of the characterisation of weak compactness for subsets of $C(X)$, stated in [DuSc58, IV.6, Theorem 14].

Theorem 13.13

Let X be a Hausdorff compact space, $H \subseteq C(X)$. Then the following properties are equivalent:

  (i)   *H is weakly relatively compact;*

  (ii) *H is bounded and relatively compact with respect to the topology $\tau_s \cap C(X)$, where τ_s denotes the product topology on $\mathbb{K}^X$;*

  (iii) *H is weakly conditionally sequentially compact.*

Proof

'(i) $\Rightarrow$ (ii)' holds, because $\tau_s \cap C(X)$ is coarser than the weak topology.

(ii) $\Rightarrow$ (iii). Let (f_n) be a sequence in H. From Theorem 13.3 we conclude that the τ_s-compact separable set $\overline{\{f_n \,;\, n \in \mathbb{N}\}}^{\tau_s} \subseteq C(X)$ is metrisable; hence there exists a τ_s-convergent subsequence (f_{n_j}). As the sequence (f_n) is bounded in $C(X)$, the dominated convergence theorem implies that (f_{n_j}) is also convergent with respect to $\sigma(C(X), \mathcal{M}(X))$.

'(iii) $\Rightarrow$ (i)' follows from Theorem 13.1. $\qquad\qquad\qquad\qquad\qquad\qquad\square$

Notes Different parts of Theorem 13.1 are due to Šmulian [Šmu40] and Eberlein [Ebe47]. The ideas of Theorem 13.3 go back to Eberlein [Ebe47] and Grothendieck [Gro52]. In fact, Grothendieck [Gro52, 3, Théorème 2] treats the case when X is only countably compact. The idea to use closures of σ-compact sets goes back to Pryce [Pry71, Theorem 2.1]. The author acknowledges substantial contributions by H. Vogt to the proof of Theorem 13.3 given above. Corollary 13.6 was shown by Eberlein for Banach spaces [Ebe47] and generalized by Grothendieck to spaces that are complete for the Mackey topology [Gro52, 4, Proposition 2]; we prove the version appearing in [Köt66, Kap. 5, § 24.2], [Sch71, Chap. IV, § 11.2]. Theorem 13.7, due to S. Dierolf, is taken from [Die78, Satz (16.1)]. The idea to use a coarser metrisable topology on E goes back to Dieudonné, Schwartz [DiSc49, section 11] and Köthe [Köt66, V, § 24.1 (3)].

Krein's Theorem

Another surprising result is Krein's theorem, stating that the closed convex hull of a weakly compact set in a Banach space is again weakly compact. This will be shown in a much more general context. For the proof, the Pettis integral of vector-valued functions will be defined and applied.

Theorem 14.1 (Krein)

Let E be a Banach space, and let $A \subseteq E$ be weakly compact. Then $\overline{\mathrm{co}}\, A$ is weakly compact.

Remarks 14.2 (a) If E is a locally convex space, $A \subseteq E$ compact, then the set $\overline{\mathrm{co}}\, A$ is precompact, hence compact if and only if it is complete (see Theorem 11.3).

This means that in the setting of Theorem 14.1 one would have to suppose that $\overline{\mathrm{co}}\, A$ is weakly complete. The "surprise" of the theorem is that the completeness of $\overline{\mathrm{co}}\, A$ in the norm topology is sufficient; see Theorem 14.8 for the more general statement.

(b) If E is a non-reflexive Banach space, then $(E, \sigma(E, E'))$ is not quasi-complete, because $\overline{B_E}^{\sigma(E'', E')} = B_{E''}$. In reflexive Banach spaces Krein's theorem is an immediate consequence of the Banach–Alaoglu theorem.

(c) In Theorem 14.1, the assertion could also have been stated in the form that '$\overline{\mathrm{aco}}\, A$ is weakly compact'. An analogous comment applies to Theorem 14.8 and Corollary 14.9. This follows from Lemma 4.9. △

Let E be a Hausdorff locally convex space, let (X, μ) be a measure space, $\mu(X) < \infty$, and let $f \colon X \to E$. Assume that $x' \circ f \in L_1(\mu)$ for all $x' \in E'$. Then $\int f \, d\mu \in E'^*$

© Springer Nature Switzerland AG 2020
J. Voigt, *A Course on Topological Vector Spaces*, Compact Textbooks in Mathematics,
https://doi.org/10.1007/978-3-030-32945-7_14

is defined by

$$\int f \, d\mu(x') := \int x' \circ f \, d\mu.$$

If additionally $\int f \, d\mu \in E$, then f is said to be **μ-Pettis-integrable**.

Remark 14.3 If X is a Hausdorff compact topological space, then, due to the Riesz–Markov representation theorem, the dual of $C(X)$ can be represented by the signed Borel measures of finite total variation on X, denoted by $\mathcal{M}(X)$. We refer to [Rud87, Theorem 2.14] for the proof of this fact. The subset of probability measures will be denoted by $\mathcal{M}_1(X)$. In what follows, the dual of $C(X)$ will simply be taken to be equal to $\mathcal{M}(X)$. The topology $\sigma(\mathcal{M}(X), C(X))$ on $\mathcal{M}(X)$ is also called the **vague topology**.

We will say that the support of a measure $\mu \in \mathcal{M}(X)$ is finite, 'spt μ finite', if μ is a linear combination of Dirac measures, i.e., there exist a finite set $F \subseteq X$ and $c_x \in \mathbb{K}$ for $x \in F$ such that $\mu = \sum_{x \in F} c_x \delta_x$, where $\delta(\{x\}) = 1$, $\delta_x(A) = 0$ for all measurable sets $A \subseteq X \setminus \{x\}$. $\hspace{2cm} \triangle$

Proposition 14.4 *Let E be a Hausdorff locally convex space. Let X be a Hausdorff compact topological space, and let $f : X \to E$ be continuous. Then the set*

$$A := \overline{\operatorname{co} f(X)}^{\sigma(E'^*, E')}$$

is $\sigma(E'^*, E')$-compact, and $A = \{\int f \, d\mu ; \ \mu \in \mathcal{M}_1(X)\}$.

Proof
The set A is compact because $f(X)$ is compact and $(E'^*, \sigma(E'^*, E'))$ is complete; recall Proposition 9.5 and Corollary 11.5(b).

It is evident that

$$\operatorname{co} f(X) = \{\textstyle\int f \, d\mu ; \ \mu \in \mathcal{M}_1(X), \ \text{spt}\,\mu \ \text{finite}\}. \tag{14.1}$$

The set $\mathcal{M}_1(X)$ is a vaguely closed subset of $B_{\mathcal{M}(X)}$, and therefore vaguely compact, by the Banach–Alaoglu theorem.

The mapping

$$\mathcal{M}(X) \ni \mu \mapsto \int f \, d\mu \in E'^*$$

is vaguely-$\sigma(E'^*, E')$-continuous. Indeed, for each $x' \in E'$, the mapping $\mathcal{M}(X) \ni \mu \mapsto \langle \int f \, d\mu, x' \rangle = \int x' \circ f \, d\mu \in \mathbb{K}$ is vaguely continuous, and as $\sigma(E'^*, E')$ is the initial topology on E'^* with respect to the mappings $E'^* \ni u \mapsto \langle u, x' \rangle \in \mathbb{K}$ $(x' \in E')$, the assertion is a consequence of Theorem 1.2. This implies that the set $\{\int f \, d\mu ; \ \mu \in \mathcal{M}_1(X)\}$

is $\sigma(E'^*, E')$-compact, and therefore (14.1) implies that

$$A \subseteq \{\textstyle\int f \, d\mu; \ \mu \in \mathcal{M}_1(X)\}.$$

On the other hand, the subspace $\{\mu \in \mathcal{M}(X); \ \mathrm{spt}\,\mu \ \text{finite}\}$ of $\mathcal{M}(X)$ separates the points of $C(X)$, which implies that this subspace is vaguely dense in $\mathcal{M}(X)$. It is not difficult to show that this implies that the positive measures of finite support are vaguely dense in the positive measures in $\mathcal{M}(X)$, and as a consequence, that the probability measures of finite support are dense in $\mathcal{M}_1(X)$. Using (14.1) again one concludes that

$$A \supseteq \{\textstyle\int f \, d\mu; \ \mu \in \mathcal{M}_1(X)\}. \qquad \square$$

Proposition 14.5 *Let E be a Hausdorff locally convex space, and let $A \subseteq E$ be compact. Then the following properties are equivalent:*
(i) *$\overline{\mathrm{co}}\, A$ is compact;*
(ii) *if X is a Hausdorff compact space, $\mu \in \mathcal{M}_1(X)$, and $f: X \to A$ is continuous, then f is μ-Pettis integrable;*
(iii) *$\overline{\mathrm{co}\, A}^{\sigma(E'^*, E')} \subseteq E$.*

Proof
'(i) $\Rightarrow$ (ii)' follows from Proposition 14.4, because (i) implies that $\overline{\mathrm{co}}\, A$ is weakly compact; therefore, $\overline{\mathrm{co}\, f(X)}^{\sigma(E'^*, E')} \subseteq \overline{\mathrm{co}\, A}^{\sigma(E'^*, E')} = \overline{\mathrm{co}\, A}^{\sigma(E, E')} \subseteq E$ for all f as in (ii).

(ii) $\Rightarrow$ (iii). With the continuous function $A \ni x \mapsto x \in E$ the assertion follows from Proposition 14.4.

(iii) $\Rightarrow$ (i). It follows from (iii) and Proposition 14.4 that $\overline{\mathrm{co}}\, A = \overline{\mathrm{co}\, A}^{\sigma(E, E')} = \overline{\mathrm{co}\, A}^{\sigma(E'^*, E')}$ is weakly compact, and therefore weakly complete. Applying Theorem 9.8 – recall that E possesses a neighbourhood base of zero consisting of (weakly) closed absolutely convex sets – we deduce that $\overline{\mathrm{co}}\, A$ is complete. As $\mathrm{co}\, A$ is also precompact, by Theorem 11.4(b), one concludes from Theorem 11.3 that $\overline{\mathrm{co}}\, A$ is compact. $\qquad \square$

Remark 14.6 A locally convex space E is said to have the **convex compactness property** if $\overline{\mathrm{co}}\, A$ is compact for every compact set $A \subseteq E$; see [Wil78, Sec. 9–2, Definition 8]. By Corollary 11.5(b), quasi-complete locally convex spaces have the convex compactness property. Proposition 14.5 can be used as a criterion for proving the convex compactness property for concrete spaces. It was used, for instance, in [Voi92] to show that the space of compact operators between two Banach spaces, equipped with the strong operator topology, has the convex compactness property (without being quasi-complete). $\qquad \triangle$

The following lemma will be needed in the proof of the general version of Krein's theorem.

Lemma 14.7 *Let E be a Hausdorff locally convex space, and let $A \subseteq E$ be weakly compact. Let $u \in \overline{\mathrm{co}\, A}^{\sigma(E'^*, E')}$, and let (x'_n) be an equicontinuous sequence in E', $x'_n \to x' \in E'$ $(n \to \infty)$ with respect to $\sigma(E', E)$.*
Then $\langle u, x'_n \rangle \to \langle u, x' \rangle$ $(n \to \infty)$.

Proof
Proposition 14.4 yields a measure $\mu \in \mathcal{M}_1(A)$ such that $u = \int_A x \, d\mu(x)$. Note that A is weakly bounded, hence bounded by Mackey's theorem, Theorem 6.1. As a consequence, the equicontinuity hypothesis implies that $\sup_{x \in A, n \in \mathbb{N}} |\langle x, x'_n \rangle| < \infty$. Therefore the convergence

$$\langle u, x'_n \rangle = \int_A \langle x, x'_n \rangle \, d\mu(x) \to \int_A \langle x, x' \rangle \, d\mu(x) = \langle u, x' \rangle \quad (n \to \infty)$$

follows from the dominated convergence theorem. □

In Lemma 14.7, if one adds the hypothesis that E is sequentially complete, then one can dispense with the hypothesis that (x'_n) is equicontinuous (see [Bou07b, IV, § 7, Exercice 10 a)]). Indeed, then the set $B := \overline{\text{aco}} A$ is a Banach disc, by Lemma 9.12(a); hence the uniform boundedness theorem, Theorem B.3, implies that the sequence (x'_n) is bounded in $(E_B)'$, and therefore $\sup_{x \in B, n \in \mathbb{N}} |\langle x, x'_n \rangle| < \infty$.

After these preparations we can show a general version of Krein's theorem.

Theorem 14.8
Let E be a Hausdorff locally convex space, and let $A \subseteq E$ be weakly compact. Then $\overline{\text{co}} \, A$ is weakly compact if and only if $\overline{\text{co}} \, A$ is complete (with respect to the topology of E).

Proof
For the necessity, we note that the hypothesis implies that $\overline{\text{co}} \, A$ is weakly complete. Using Theorem 9.8 we conclude that $\overline{\text{co}} \, A$ is complete.

For the sufficiency, we first show the assertion under the additional assumption that E is separable. Let $\tilde{E}$ be the completion of E. As $\overline{\text{co}} \, A$ is complete, we conclude that $\overline{\text{co}} \, A = \overline{\text{co}} \, A^{\tilde{E}}$, and therefore the assertion is equivalent to $\overline{\text{co}} \, A^{\tilde{E}}$ being $\sigma(\tilde{E}, E')$-compact. This means that without loss of generality we may assume that E is complete.

Let $u \in \overline{\text{co}} \, A^{\sigma(E'^*, E')}$. Let $M \subseteq E'$ be equicontinuous and $\sigma(E', E)$-closed. Then M is compact, by the Alaoglu–Bourbaki theorem, therefore metrisable, because E is separable; cf. Proposition 4.11. Now Lemma 14.7 shows that $u|_M$ is (sequentially) $\sigma(E', E)$-continuous. Therefore, Corollary 9.16 shows that $u \in E$.

This shows that $\overline{\text{co}} \, A^{\sigma(E'^*, E')} \subseteq E$, and Proposition 14.5 (applied in the space $(E, \sigma(E, E'))$) implies that $\overline{\text{co}} \, A = \overline{\text{co}} \, A^{\sigma(E, E')}$ is weakly compact.

For the general case we recall from Corollary 13.6 that it is sufficient to show that co A is weakly conditionally countably compact. Let (x_n) be a sequence in co A. Then there exists a closed separable subspace E_0 of E such that the sequence (x_n) belongs to $\text{co}(A \cap E_0)$; for convenience we introduce $A_0 := A \cap E_0$. Then it is clear that the set $\overline{\text{co}} \, A_0 \subseteq (\overline{\text{co}} \, A) \cap E_0$ is closed and therefore complete in E_0. Now the separable case treated above shows that (x_n) possesses a weak cluster point in E_0, which then is also a weak cluster point in E. □

Proof of Theorem 14.1

Krein's theorem is a special case of Theorem 14.8.   □

In fact, Theorem 14.8 shows that in Theorem 14.1 one can replace 'Banach space' by 'quasi-complete Hausdorff locally convex space'.

In Theorem 14.8, the topology of E can be chosen as any topology compatible with the dual pair $\langle E, E' \rangle$; in fact, the weakest form of the completeness hypothesis is to assume that $\overline{\text{co}}\, A$ is $\mu(E, E')$-complete. The following generalisation shows that the weak topology can be replaced by other compatible topologies.

Corollary 14.9 *Let (E, ρ) be a Hausdorff locally convex space, and let $A \subseteq E$ be ρ-compact. Then $\overline{\text{co}}\, A$ is ρ-compact if and only if $\overline{\text{co}}\, A$ is complete for the Mackey topology $\mu(E, E')$.*

Proof

We apply Theorem 14.8 with E equipped with the topology $\tau := \mu(E, E')$; recall that the τ-closure of co A is equal to the ρ- and $\sigma(E, E')$-closure, by Mazur's theorem, Corollary 2.11.

As A is ρ-compact, co A is ρ-precompact. Therefore the set $\overline{\text{co}}\, A$ is ρ-compact if and only if it is ρ-complete.

Now, if $\overline{\text{co}}\, A$ is ρ-complete, then it is τ-complete, by Theorem 9.8.

If the set $\overline{\text{co}}\, A$ is τ-complete, then it is $\sigma(E, E')$-compact by Theorem 14.8 (note that the ρ-compactness of A implies the $\sigma(E, E')$-compactness), therefore $\sigma(E, E')$-complete, and then ρ-complete by Theorem 9.8.   □

Notes Theorem 14.1 is due to M. Krein [Kre37] for separable E and to Krein and Šmulian [KrŠm40, Theorem 24] for the general case. Proposition 14.4 is essentially from Bourbaki [Bou07b, III, § 3, Proposition 5], and Proposition 14.5 uses the ideas in [Bou07b, III, § 3 and IV, § 7]. Lemma 14.7 is adapted from [Bou07b, IV, § 7, Exercice 10 a)], and our proof of Theorem 14.8 is given as suggested in [Bou07b, IV, § 7, Exercice 10]. Corollary 14.9 is taken from Schaefer [Sch71, Chap. IV, § 11.5].

Weakly Compact Sets in $L_1(\mu)$

In view of the discussion of properties of weakly compact sets in the last chapters, it seems appropriate to present examples of weakly compact sets in a non-reflexive space. Besides the characterisation of weak compactness of subsets of $L_1(\mu)$, we will also show that $L_1(\mu)$ is weakly sequentially complete.

In all of this chapter $(\Omega, \mathcal{A}, \mu)$ will be a measure space.

A set $H \subseteq L_1(\mu)$ is called **equi-integrable** if H is bounded and for any sequence (A_n) in $\mathcal{A}$ with $A_n \supseteq A_{n+1}$ for all $n \in \mathbb{N}$ and $\bigcap_{n\in\mathbb{N}} A_n = \varnothing$, one has $\sup_{f\in H} \int_{A_n} |f| \, d\mu \to 0$ as $n \to \infty$.

The main objective of this chapter is to prove the Dunford–Pettis theorem, which asserts that weak relative compactness for a subset of $L_1(\mu)$ is equivalent to equi-integrability; see Theorem 15.4.

We warn the reader that the notion of equi-integrability (also sometimes called 'uniform integrability') in some references is defined without the requirement of boundedness, and quite generally, there are various definitions of equi-integrability around, not all equivalent.

For functions $f, g \colon \Omega \to \mathbb{R}$ we will use the notation $[f > g] := \{x \in \Omega;\; f(x) > g(x)\}$, and similarly for $[f > 0]$, etc. In order to obtain another formulation of equi-integrability where in the condition the terms $\int_{A_n} |f| \, d\mu$ are replaced by $\left| \int_{A_n} f \, d\mu \right|$, we make the following observation. For $f \in L_1(\mu)$, $A \in \mathcal{A}$ there exists $B \in \mathcal{A}$, $B \subseteq A$ such that $\int_A |f| \, d\mu \leqslant 4 \left| \int_B f \, d\mu \right|$. To show this we first observe that $\int_A |f| \, d\mu \leqslant \int_A |\operatorname{Re} f| \, d\mu + \int_A |\operatorname{Im} f| \, d\mu$, and without loss of generality we can assume that $\int_A |\operatorname{Im} f| \, d\mu \leqslant \int_A |\operatorname{Re} f| \, d\mu$. Let $A_\pm := [\pm \operatorname{Re} f > 0]$; also without loss of generality we may assume that $-\int_{A_-} \operatorname{Re} f \, d\mu \leqslant \int_{A_+} \operatorname{Re} f \, d\mu$. Then with $B := A_+$ one obtains

$$\int_A |f| \, d\mu \leqslant 2 \int_A |\operatorname{Re} f| \, d\mu \;\leqslant\; 4 \int_B \operatorname{Re} f \, d\mu \leqslant 4 \left| \int_B f \, d\mu \right|.$$

J. Voigt, *A Course on Topological Vector Spaces*, Compact Textbooks in Mathematics,
https://doi.org/10.1007/978-3-030-32945-7_15

In what follows we will use the abbreviation $A_n \downarrow \varnothing$ for a decreasing sequence (A_n) of sets satisfying $\bigcap_n A_n = \varnothing$.

Lemma 15.1 *A set $H \subseteq L_1(\mu)$ is equi-integrable if and only if it is bounded and for all sequences (A_n) in $\mathcal{A}$ with $A_n \downarrow \varnothing$ one has $\sup_{f \in H} \left| \int_{A_n} f \, \mathrm{d}\mu \right| \to 0$ as $n \to \infty$.*

Proof

It is trivial that the equi-integrability of H implies the condition. The converse implication will be proved by contraposition. Thus, assume that H is not equi-integrable. It is not difficult to show that then there exist $\varepsilon > 0$, a sequence (B_n) in $\mathcal{A}$, $B_n \downarrow \varnothing$, and a sequence (f_n) in H such that $\int_{B_n} |f_n| \, \mathrm{d}\mu \geqslant \frac{9}{8}\varepsilon$ and $\int_{B_{n+1}} |f_n| \, \mathrm{d}\mu \leqslant \frac{\varepsilon}{8}$ for all $n \in \mathbb{N}$. Note that this implies that $\int_{B_n \setminus B_{n+1}} |f_n| \, \mathrm{d}\mu \geqslant \varepsilon$ for all $n \in \mathbb{N}$. Then, by the observation preceding the lemma, for each $n \in \mathbb{N}$ there exists a set $C_n \in \mathcal{A}$, $C_n \subseteq B_n \setminus B_{n+1}$ such that $\left| \int_{C_n} f_n \, \mathrm{d}\mu \right| \geqslant \frac{1}{4} \int_{B_n \setminus B_{n+1}} |f_n| \, \mathrm{d}\mu \geqslant \frac{\varepsilon}{4}$. Defining $A_n := \bigcup_{k=n}^{\infty} C_k$ $(n \in \mathbb{N})$ we obtain $A_n \downarrow \varnothing$ and

$$\left| \int_{A_n} f_n \, \mathrm{d}\mu \right| \geqslant \left| \int_{A_n \setminus A_{n+1}} f_n \, \mathrm{d}\mu \right| - \int_{A_{n+1}} |f_n| \, \mathrm{d}\mu$$

$$\geqslant \left| \int_{C_n} f_n \, \mathrm{d}\mu \right| - \int_{B_{n+1}} |f_n| \, \mathrm{d}\mu \geqslant \frac{\varepsilon}{4} - \frac{\varepsilon}{8} = \frac{\varepsilon}{8};$$

hence, $\sup_{f \in H} \left| \int_{A_n} f \, \mathrm{d}\mu \right| \geqslant \frac{\varepsilon}{8}$ for all $n \in \mathbb{N}$. $\square$

In the proof that equi-integrability implies weak relative compactness we will use the following weak compactness criterion for sets in Banach spaces.

Lemma 15.2 (Grothendieck) *Let E be a Banach space, and let $A \subseteq E$. Assume that for all $\varepsilon > 0$ there exists a weakly compact set $A_\varepsilon \subseteq E$ such that $A \subseteq A_\varepsilon + \varepsilon B_E$. Then A is weakly relatively compact.*

Proof

Obviously A is bounded, and therefore $\overline{A}^{\sigma(E'', E')}$ is $\sigma(E'', E')$-compact. It is sufficient to show that $\overline{A}^{\sigma(E'', E')} \subseteq E$. For $\varepsilon > 0$ one has

$$\overline{A}^{\sigma(E'', E')} \subseteq \overline{A_\varepsilon + \varepsilon B_E}^{\sigma(E'', E')} \subseteq A_\varepsilon + \varepsilon B_{E''},$$

where for the last inclusion we have used that $A_\varepsilon + \varepsilon B_{E''}$ is $\sigma(E'', E')$-compact. This implies that

$$\overline{A}^{\sigma(E'', E')} \subseteq \bigcap_{\varepsilon > 0} (A_\varepsilon + \varepsilon B_{E''}).$$

Given $x \in \overline{A}^{\sigma(E'', E')}$, one obtains sequences (x_n) in E, (y_n) in E'', $x_n \in A_{1/n}$, $\|y_n\| \leqslant 1/n$, $x_n + y_n = x$ $(n \in \mathbb{N})$. From $y_n \to 0$ one concludes that $x_n \to x$ $(n \to \infty)$, hence $x \in E$. $\square$

In order to apply this criterion we have to deduce the required ε-approximation from the equi-integrability. This will be provided by the following lemma.

Lemma 15.3 *Let $H \subseteq L_1(\mu)$ be equi-integrable. Then:*

(a) *For any $\varepsilon > 0$ there exists $\delta > 0$ such that $B \in \mathcal{A}$, $\mu(B) < \delta$ implies that $\int_B |f| \, d\mu \leqslant \varepsilon$ for all $f \in H$.*

(b) *For any $\varepsilon > 0$ there exists $B \in \mathcal{A}$ with $\mu(B) < \infty$ such that $\int_{\Omega \setminus B} |f| \, d\mu \leqslant \varepsilon$ for all $f \in H$.*

(c) *For any $\varepsilon > 0$ there exist $B \in \mathcal{A}$ with $\mu(B) < \infty$ and $n \in \mathbb{N}$ such that $\sup_{f \in H} \int (|f| - n\mathbf{1}_B)^+ \, d\mu \leqslant \varepsilon$.*

Proof

(a) Assume that the assertion does not hold. Then there exists $\varepsilon > 0$ such that for all $n \in \mathbb{N}$ one can find a set $B_n \in \mathcal{A}$ with $\mu(B_n) \leqslant 2^{-n}$ and $f_n \in H$ such that $\int_{B_n} d\mu \geqslant \varepsilon$. Then $B_0 := \bigcap_{n \in \mathbb{N}} \bigcup_{k=n}^{\infty} B_k$ is a μ-null set, and setting $A_n := \left(\bigcup_{k=n}^{\infty} B_k \right) \setminus B_0$ one obtains a sequence (A_n) in $\mathcal{A}$ such that $A_n \downarrow \varnothing$ and $\int_{A_n} |f_n| \, d\mu \geqslant \int_{B_n} |f_n| \, d\mu \geqslant \varepsilon$ for all $n \in \mathbb{N}$, which contradicts the equi-integrability of H.

(b) Assume that the assertion does not hold. Then there exists $\varepsilon > 0$ such that for all $B \in \mathcal{A}$ with $\mu(B) < \infty$ one can find $f \in H$ such that $\int_{\Omega \setminus B} |f| \, d\mu > \varepsilon$. This implies that there exist a disjoint sequence (B_n) in $\mathcal{A}$ with $\mu(B_n) < \infty$ for all $n \in \mathbb{N}$ and a sequence (f_n) in H such that $\int_{B_n} |f_n| \, d\mu \geqslant \varepsilon$ for all $n \in \mathbb{N}$. Setting $A_n := \bigcup_{k \geqslant n} B_k$ we obtain a sequence (A_n) in $\mathcal{A}$, $A_n \downarrow \varnothing$, with $\int_{A_n} |f_n| \, d\mu \geqslant \varepsilon$ for all $n \in \mathbb{N}$, which contradicts the equi-integrability of H.

(c) Let $\varepsilon > 0$. Because of part (b) above, there exists $B \in \mathcal{A}$ with $\mu(B) < \infty$ such that

$$\int_{\Omega \setminus B} |f| \, d\mu \leqslant \varepsilon/2 \quad (f \in H). \tag{15.1}$$

Define $c := \sup_{f \in H} \|f\|$. By part (a), there exists $\delta > 0$ such that $\int_A |f| \, d\mu < \varepsilon/2$ for all $f \in H$ and all $A \in \mathcal{A}$ with $\mu(A) < \delta$. For $n \in \mathbb{N}$, $f \in H$ we obtain

$$c \geqslant \int_{[|f| > n]} |f| \, d\mu \geqslant n\mu([|f| > n]).$$

For $n > c/\delta$ we conclude that $\mu([|f| > n]) \leqslant c/n < \delta$; hence

$$\|(|f| - n)^+\| = \int_{[|f| > n]} (|f| - n) \, d\mu < \varepsilon/2 \quad (f \in H, \ n > c/\delta). \tag{15.2}$$

Combining (15.1) and (15.2) we obtain the assertion. $\qquad \square$

We mention in passing that in fact a set $H \subseteq L_1(\mu)$ is equi-integrable if and only if H is bounded and the properties asserted in (a) and (b) of Lemma 15.3 are satisfied. Another noteworthy consequence of part (b) is that an equi-integrable set H always 'lives on a σ-finite subset of Ω': There exists a σ-finite subset $B \in \mathcal{A}$ such that $f|_{\Omega \setminus B} = 0$ for all $f \in H$.

For one of the equivalences in the main result of this chapter we introduce the following notation. For a disjoint sequence (B_n) in $\mathcal{A}$ we define the mapping

$$L_{(B_n)} \colon L_1(\mu) \to \ell_1, \ f \mapsto \left(\int_{B_n} f \, d\mu \right)_{n \in \mathbb{N}}.$$

Obviously $L_{(B_n)}$ is a continuous linear operator, even a contraction.

Theorem 15.4 (Dunford–Pettis)

For $H \subseteq L_1(\mu)$ the following properties are equivalent:
  (i)   *H is equi-integrable;*
  (ii) *H is weakly relatively compact;*
  (iii) *for each disjoint sequence (B_n) in $\mathcal{A}$ the operator $L_{(B_n)}$ maps H to a relatively compact subset of ℓ_1.*

Proof

(i) $\Rightarrow$ (ii). Let $\varepsilon > 0$. We choose B and n as asserted in Lemma 15.3(c). In $L_2(B, \mu_B)$, where μ_B denotes the restriction of the measure μ to $\mathcal{A} \cap B$, the set $\{ f \in L_2(\mu_B); \ |f| \leqslant n \}$ is bounded, convex and closed, hence weakly compact (because $L_2(\mu_B)$ is reflexive). The embedding $L_2(\mu_B) \hookrightarrow L_1(\mu_B)$ is continuous, hence, by Lemma 6.3, continuous with respect to the weak topologies, and as a consequence the set $\{ f \in L_1(\mu); \ |f| \leqslant n \mathbf{1}_B \}$ is weakly compact in $L_1(\mu)$. The inequality in Lemma 15.3(c) shows that $H \subseteq \{ f \in L_1(\mu); \ |f| \leqslant n \mathbf{1}_B \} + B_{L_1(\mu)}(0, \varepsilon)$. Now Lemma 15.2 implies that H is weakly relatively compact.

  (ii) $\Rightarrow$ (iii). As $L_{(B_n)} \colon L_1(\mu) \to \ell_1$ is a continuous operator, this operator is also continuous with respect to the weak topologies; hence $L_{(B_n)}(H)$ is a weakly relatively compact subset of ℓ_1, and Corollary 5.10 implies that $L_{(B_n)}(H)$ is relatively compact.

  (iii) $\Rightarrow$ (i). Let (A_n) be a sequence in $\mathcal{A}$, $A_n \downarrow \varnothing$. We define $B_n := A_n \setminus A_{n+1}$ $(n \in \mathbb{N})$. Then (B_n) is a disjoint sequence in $\mathcal{A}$.

  Clearly, $L_{(B_n)}(H)$ is bounded. Recall from Example 5.6(i) that the relative compactness of $L_{(B_n)}(H)$ is equivalent to $\sup_{f \in H} \sum_{k=n}^{\infty} \left| \int_{B_k} f \, d\mu \right| \to 0$ as $n \to \infty$. Observe that

$$\left| \int_{A_n} f \, d\mu \right| \leqslant \sum_{k=n}^{\infty} \left| \int_{B_k} f \, d\mu \right| \qquad (f \in H, \ n \in \mathbb{N}).$$

Hence, Lemma 15.1 implies that H is equi-integrable.   □

As the second important result of the present chapter we show that $L_1(\mu)$ is weakly sequentially complete. For ℓ_1, this property had already been shown in Theorem 5.8; see also Remark 5.9.

Let (f_n) be a Cauchy sequence in $L_1(\mu)$ with respect to the weak topology. Then (f_n) is weakly convergent.

Proof

Let (B_k) be a disjoint sequence in $\mathcal{A}$. Then it is immediate that $(L_{(B_k)} f_n)_n$ is a weak Cauchy sequence in ℓ_1; recall Lemma 6.3. Theorem 5.8 implies that $(L_{(B_k)} f_n)_n$ is convergent, and therefore the range of the sequence is relatively compact in ℓ_1.

Now Theorem 15.4 shows that the set $\{f_n ;\, n \in \mathbb{N}\}$ is weakly relatively compact. This implies that the sequence (f_n) possesses a weak cluster point. Being a weak Cauchy sequence, it is convergent in the weak topology, by Remark 9.1(b).　　　□

We conclude this chapter by some additional comments.

Remarks 15.6 (a) It is not difficult to show that the equi-integrability of a set $H \subseteq L_1(\mu)$ is equivalent to the condition that for each $\varepsilon > 0$ there exists $g \in L_1(\mu)_+$ such that $\sup_{f \in H} \int_{[|f| > g]} |f| \, d\mu < \varepsilon$.
Concerning the necessity of this condition, the function g can be found in the form $g = c\mathbf{1}_B$ for suitable $c > 0$ and $B \in \mathcal{A}$ with $\mu(B) < \infty$; see Lemma 15.3(c). The sufficiency is rather immediate.

(b) Theorem 15.4 implies: If $H \subseteq L_1(\mu)$ is weakly relatively compact, then the set

$$\{f \in L_1(\mu);\ \text{there exists } g \in H \text{ such that } |f| \leqslant |g|\}$$

is weakly relatively compact. In particular, for every $g \in L_1(\mu)_+$ the order interval

$$[-g, g] := \{f \in L_1(\mu);\ -g \leqslant f \leqslant g\}$$

is weakly compact.

Similarly: If $g \in L_1(\mathbb{R})_+$, then the set $\{g(\cdot - y);\ 0 \leqslant y \leqslant 1\}$ is compact (because the mapping $y \mapsto g(\cdot - y)$ is continuous); hence,

$$\{f \in L_1(\mu);\ |f| \leqslant g(\cdot - y) \text{ for some } y \in [0, 1]\}$$

is weakly relatively compact.　　　△

Notes Theorem 15.4 is due to Dunford and Pettis [DuPe40, Theorem 3.2.1]. Lemma 15.2 is attributed to Grothendieck in [Die84, XIII, Lemma 2]. With the aid of this lemma the proof that equi-integrability implies weak relative compactness, in Theorem 15.4, is rather natural. The author was at a loss for finding a short 'measure theory-free' proof of the reverse implication, in the literature. The device to use the operators $L_{(B_n)}$ in Theorem 15.4 is present in the original paper [DuPe40], for 'decompositions' of the measure space. The weak sequential completeness of $L_1(\mu)$, Theorem 15.5, is due to Dunford and Pettis as well [DuPe40, p. 377].

Our definition of equi-integrability can be found implicitly in [DuSc58, Theorem IV.8.9, Corollary IV.8.10 and their proofs]. The characterisation of equi-integrability mentioned in Remark 15.6(a) is taken as the definition in [Bau90, § 21] and appears in [Bog07, Theorem 4.7.20] as one of the equivalences of weak relative compactness of a set.

$$\mathcal{B}_0'' = \mathcal{B}$$

The issue of this chapter is to present an example where one can compute the bidual of a locally convex space without having an explicit description of the dual. This example could have been inserted much earlier, in fact after Chapter 8. We have preferred, however, to first pursue more theoretical developments.

The spaces mentioned in the title of the chapter are

$$\mathcal{B}_0 = \mathcal{B}_0(\mathbb{R}^n) := C_0^\infty(\mathbb{R}^n), \quad \mathcal{B} = \mathcal{B}(\mathbb{R}^n) := C_b^\infty(\mathbb{R}^n),$$

provided with the norms p_m ($m \in \mathbb{N}_0$),

$$p_m(f) := \sup_{x \in \mathbb{R}^n, |\alpha| \leqslant m} |\partial^\alpha f(x)| = \max_{|\alpha| \leqslant m} \|\partial^\alpha f\|_\infty.$$

These spaces are Fréchet spaces, and $\mathcal{B}_0$ is the closure of $\mathcal{D}(\mathbb{R}^n)$ in $\mathcal{B}(\mathbb{R}^n)$.

We recall from Theorem 3.2 how the bidual of a Hausdorff locally convex space can be obtained as a subset of E'^* (with the neighbourhood base of zero $\mathcal{U}' := \{B^\circ;\ B \subseteq E \text{ bounded}\}$ of $(E', \beta(E', E))$, and the polars in the dual pair $\langle E'^*, E' \rangle$ denoted by '•'):

$$E'' = (E', \beta(E', E))' = \bigcup_{U \in \mathcal{U}'} U^\bullet = \bigcup_{B \subseteq E \text{ bounded}} B^{\circ\bullet}$$

$$= \bigcup_{B \subseteq E \text{ bounded}} \overline{\mathrm{aco}\, B}^{\,\sigma(E'^*, E')}. \tag{16.1}$$

Our aim is to show that $\mathcal{B}_0'' = \mathcal{B}$ (in a suitable interpretation). We will do this in four steps, as follows:

1. Determine the bounded sets of $\mathcal{B}_0$ and $\mathcal{B}$.
2. Show a continuity property of the elements of $\mathcal{B}_0'$, allowing to extend them to $\mathcal{B}$.

J. Voigt, *A Course on Topological Vector Spaces*, Compact Textbooks in Mathematics,
https://doi.org/10.1007/978-3-030-32945-7_16

3. Embed $\mathcal{B}$ into $\mathcal{B}_0'^*$, determine the $\sigma(\mathcal{B}_0'^*, \mathcal{B}_0')$-closures of bounded sets of $\mathcal{B}_0$, and show that $\mathcal{B}_0'' = \mathcal{B}$ as sets.
4. Show that $\beta(\mathcal{B}, \mathcal{B}_0')$ is the Fréchet space topology of $\mathcal{B}$.

Ad 1. A set $A \subseteq \mathcal{B}_0$ is bounded if and only if $\sup_{f \in A} p_k(f) < \infty$ for all $k \in \mathbb{N}$. This means that for any sequence $c = (c_k)_k \in (0, \infty)^{\mathbb{N}_0}$ the set

$$A_c := \left\{ f \in \mathcal{B}_0; \ p_k(f) \leqslant c_k \ (k \in \mathbb{N}) \right\}$$

is bounded. Therefore, the collection

$$\mathcal{M} := \left\{ A_c; \ c \in (0, \infty)^{\mathbb{N}_0} \right\}.$$

is a cobase of bounded subsets of $\mathcal{B}_0$, and this implies that $\beta(\mathcal{B}_0', \mathcal{B}_0) = \tau_{\mathcal{M}}$, and that

$$\left\{ A_c^{\circ}; \ c \in (0, \infty)^{\mathbb{N}_0} \right\}$$

is a neighbourhood base of zero for $\beta(\mathcal{B}_0', \mathcal{B}_0)$.

Analogously, a set $\hat{A} \subseteq \mathcal{B}$ is bounded if and only if $\hat{A} \subseteq \hat{A}_c := \left\{ f \in \mathcal{B}; \ p_k(f) \leqslant c_k \ (k \in \mathbb{N}) \right\}$ for some sequence $c = (c_k)$ in $(0, \infty)$.

Ad 2. We note that the embedding $\mathcal{D} \hookrightarrow \mathcal{B}_0$ (with $\mathcal{D} := \mathcal{D}(\mathbb{R}^n)$) is dense and continuous. This implies that $\mathcal{B}_0' \subseteq \mathcal{D}'$, in the sense that $u|_{\mathcal{D}} \in \mathcal{D}'$ for all $u \in \mathcal{B}_0'$.

Lemma 16.1 *Let $u \in \mathcal{B}_0'$. Then there exist $m \in \mathbb{N}_0$, $C \geqslant 0$ such that*

$$|\langle f, u \rangle| \leqslant C p_m(f) \quad (f \in \mathcal{B}_0).$$

For all $\varepsilon > 0$ there exists a compact set $K \subseteq \mathbb{R}^n$ such that for all $\varphi \in \mathcal{D}$ with $\mathrm{spt}\, \varphi \cap K = \varnothing$ one has

$$|\langle \varphi, u \rangle| \leqslant \varepsilon p_m(\varphi).$$

Proof
The first statement is just the continuity of u.

Assume that the second property does not hold. Then there exist $\varepsilon > 0$ and a sequence (φ_k) in $\mathcal{D}$ with disjoint supports, $p_m(\varphi_k) = 1$, $\langle \varphi_k, u \rangle > \varepsilon$ $(k \in \mathbb{N})$. Then $\sum_{j=1}^{k} \varphi_j \in \mathcal{D}$, $p_m(\sum_{j=1}^{k} \varphi_j) = 1$, $\langle \sum_{j=1}^{k} \varphi_j, u \rangle > k\varepsilon$ $(k \in \mathbb{N})$. Because of $k\varepsilon \to \infty$ $(k \to \infty)$ one obtains a contradiction to the first inequality. $\qquad \square$

For all $m \in \mathbb{N}_0$ there exists $M_m \geqslant 0$ such that

$$p_m(fg) \leqslant M_m p_m(f) p_m(g)$$

for all $f, g \in \mathcal{B}$; a consequence of the product rule.

Let $(\eta_k)_{k \in \mathbb{N}}$ be a **special approximate unit**, i.e., (η_k) is a sequence in $\mathcal{D}$ which is bounded in $\mathcal{B}$, and for each compact set $K \subseteq \mathbb{R}^n$ one has that $\eta_k|_K = 1$ for large k.

Lemma 16.2 *Let $u \in \mathcal{B}'_0$. Then for $f \in \mathcal{B}$ the limit*

$$\langle f, u \rangle^* := \lim_{k \to \infty} \langle \eta_k f, u \rangle$$

exists. If $\hat{A} \subseteq \mathcal{B}$ is bounded, then

$$\langle \eta_k f, u \rangle \to \langle f, u \rangle^* \quad (k \to \infty),$$

uniformly in $f \in \hat{A}$.

Proof

Let $m \in \mathbb{N}_0$ be as in Lemma 16.1. Let $\varepsilon > 0$. Then for

$$\varepsilon' := \frac{\varepsilon}{2 M_m \, p_m(f) \, \sup_{k \in \mathbb{N}} p_m(\eta_k)}$$

there exists a compact set $K \subseteq \mathbb{R}^n$ such that for all $\varphi \in \mathcal{D}$ with $\mathrm{spt}\, \varphi \cap K = \varnothing$ one has $|\langle \varphi, u \rangle| \leqslant \varepsilon' p_m(\varphi)$. By the properties of (η_k), there exists k_0 such that $\eta_k = 1$ in a neighbourhood of K, for $k \geqslant k_0$. For $k, l \geqslant k_0$ one therefore has $(\eta_k - \eta_l) f \in \mathcal{D}$, $\mathrm{spt}\big((\eta_k - \eta_l) f\big) \cap K = \varnothing$, hence

$$|\langle \eta_k f, u \rangle - \langle \eta_l f, u \rangle| = |\langle (\eta_k - \eta_l) f, u \rangle|$$

$$\leqslant \varepsilon' p_m((\eta_k - \eta_l) f) \leqslant \varepsilon' M_m \, p_m(\eta_k - \eta_l) \, p_m(f) \leqslant \varepsilon.$$

This shows the existence of $\langle f, u \rangle^*$.

There exists $c \in (0, \infty)^{\mathbb{N}_0}$ such that $\hat{A} \subseteq \hat{A}_c$. Put

$$\varepsilon' := \frac{\varepsilon}{2 M_m c_m \, \sup_{k \in \mathbb{N}} p_m(\eta_k)}.$$

With the compact set $K \subseteq \mathbb{R}^n$ and k_0 chosen as above, one deduces that

$$|\langle \eta_k f, u \rangle - \langle f, u \rangle^*| \leqslant \varepsilon' M_m 2 \sup_{k \in \mathbb{N}} p_m(\eta_k) \, p_m(f) \leqslant \varepsilon$$

for all $f \in \hat{A}_c$, $k \geqslant k_0$. $\qquad \square$

Remark 16.3 Let $u \in \mathcal{B}'_0$. Then there exist m, C as in Lemma 16.1, which means that $u \in (\mathcal{B}_0, p_m)'$. The mapping

$$j : (\mathcal{B}_0, p_m) \ni f \mapsto (\partial^\alpha f)_{|\alpha| \leqslant m} \in C_0(\mathbb{R}^n)^{\{\alpha;\ |\alpha| \leqslant m\}} =: E_m$$

is isometric. The Hahn–Banach theorem implies that there exists $\hat{u} \in E_m'$ such that $u = \hat{u} \circ j$. Since $C_0(\mathbb{R}^n)' = \mathcal{M}_f(\mathbb{R}^n)$ (Borel measures of finite total variation; the Riesz–Markov representation theorem), there exist $\mu_\alpha \in \mathcal{M}_f(\mathbb{R}^n)$, for $|\alpha| \leqslant m$, such that

$$\langle f, u \rangle = \sum_{|\alpha| \leqslant m} \int_{\mathbb{R}^n} \partial^\alpha f \, d\mu_\alpha = \sum_{|\alpha| \leqslant m} (-1)^{|\alpha|} \langle f, \partial^\alpha \mu_\alpha \rangle$$

$$= \langle f, \sum_{|\alpha| \leqslant m} (-1)^{|\alpha|} \partial^\alpha \mu_\alpha \rangle \qquad (f \in \mathcal{B}_0),$$

where the derivatives on the measures are to be interpreted in the distributional sense. In particular, one then obtains

$$\langle \mathbf{1}, u \rangle^* = \int d\mu_0.$$

The elements of $\mathcal{B}_0'$ are called **integrable distributions**. $\qquad \triangle$

Ad 3. Evidently, $\langle \cdot, \cdot \rangle^*$ is a bilinear form on $\mathcal{B} \times \mathcal{B}_0'$ and therefore induces a linear mapping $\hat{\kappa} : \mathcal{B} \to \mathcal{B}_0'^*$, $\hat{\kappa}(f)(u) := \langle f, u \rangle^*$.

First we show that $\langle f, u \rangle^* = \langle f, u \rangle$ for all $f \in \mathcal{B}_0, u \in \mathcal{B}_0'$. Indeed, $\eta_k f \to f$ in $\mathcal{B}_0$, because

$$\|\partial^\alpha (\eta_k - 1) f\|_\infty \leqslant \sum_{0 \leqslant \beta \leqslant \alpha} \binom{\alpha}{\beta} \|\partial^{\alpha-\beta}(\eta_k - 1)\|_\infty \sup_{x \in \operatorname{spt}(\eta_k - 1)} |\partial^\beta f(x)| \to 0$$

as $k \to \infty$, for all $\alpha \in \mathbb{N}_0^n$. Therefore $\langle \eta_k f, u \rangle \to \langle f, u \rangle$ $(k \to \infty)$. This shows that $\hat{\kappa}$ is an extension of the canonical embedding $\kappa : \mathcal{B}_0 \hookrightarrow \mathcal{B}_0'^*$.

Also, $\hat{\kappa}$ is injective. Indeed, let $f \in \mathcal{B}$ such that $\langle f, u \rangle^* = 0$ for all $u \in \mathcal{B}_0'$. As the evaluation functionals δ_x are elements of $\mathcal{B}_0'$, one obtains

$$0 = \langle f, \delta_x \rangle^* = f(x) \quad (x \in \mathbb{R}^n),$$

i.e., $f = 0$. Now that we have embedded $\mathcal{B}$ into $\mathcal{B}_0'^*$ we can consider the dual pair $\langle \mathcal{B}, \mathcal{B}_0' \rangle$ with the bilinear form $\langle \cdot, \cdot \rangle^*$.

In the following we denote by ρ the standard topology of $\mathcal{E} = \mathcal{E}(\mathbb{R}^n)$ $(= C^\infty(\mathbb{R}^n))$, provided with the topology of compact convergence of all derivatives. Let $c \in (0, \infty)^{\mathbb{N}_0}$. As $\mathcal{B} \hookrightarrow \mathcal{E}$ is continuous, the set $\hat{A}_c$ is bounded (and closed) in $\mathcal{E}$, and therefore $\hat{A}_c$ is compact in $\mathcal{E}$, because $\mathcal{E}$ is a Montel space; see Chapter 8.

Lemma 16.4 *Let $u \in \mathcal{B}_0'$, $c \in (0, \infty)^{\mathbb{N}_0}$. Then the mapping*

$$\hat{A}_c \ni f \mapsto \langle f, u \rangle^* \in \mathbb{K}$$

is continuous with respect to the topology ρ.

Proof

For $k \in \mathbb{N}$ the mapping $f \mapsto \langle \eta_k f, u \rangle$ is continuous with respect to ρ (because $\mathcal{E} \ni f \mapsto \eta_k f \in \mathcal{B}_0$ is continuous). By Lemma 16.2, these mappings converge uniformly on $\hat{A}_c$ to $\langle f, u \rangle^*$. This implies the assertion. $\qquad \square$

Theorem 16.5

Let $c \in (0, \infty)^{\mathbb{N}_0}$. Then:

(a) *$\hat{A}_c$ is $\sigma(\mathcal{B}, \mathcal{B}_0')$-compact, and $\overline{A_c}^{\sigma(\mathcal{B}_0'^*, \mathcal{B}_0')} \subseteq \hat{A}_c$.*

(b) *There exists $d \in (0, \infty)^{\mathbb{N}_0}$ such that $\hat{A}_c \subseteq \overline{A_d}^{\sigma(\mathcal{B}_0'^*, \mathcal{B}_0')}$.*

(c) *$\mathcal{B}_0'' = \mathcal{B}$ (as sets).*

Proof

(a) For $u \in \mathcal{B}_0'$ the mapping $\hat{A}_c \ni f \mapsto \langle f, u \rangle^*$ is continuous with respect to $\rho \cap \hat{A}_c$, by Lemma 16.4. As $\sigma(\mathcal{B}, \mathcal{B}_0') \cap \hat{A}_c$ is the initial topology with respect to the mappings $f \mapsto \langle f, u \rangle^*$ ($u \in \mathcal{B}_0'$), one obtains the continuity of id: $(\hat{A}_c, \rho \cap \hat{A}_c) \to (\hat{A}_c, \sigma(\mathcal{B}, \mathcal{B}_0') \cap \hat{A}_c)$. Therefore, the ρ-compactness of $(\hat{A}_c, \rho \cap \hat{A}_c)$ implies the $\sigma(\mathcal{B}, \mathcal{B}_0')$-compactness of $\hat{A}_c$, which in turn implies that $\hat{A}_c$ is $\sigma(\mathcal{B}_0'^*, \mathcal{B}_0')$-closed. Then from $A_c \subseteq \hat{A}_c$ one obtains the asserted inclusion.

(b) For $f \in \hat{A}_c$, $m \in \mathbb{N}_0$ one has

$$p_m(\eta_k f) \leqslant M_m \sup_{k \in \mathbb{N}} p_m(\eta_k) c_m =: d_m.$$

Therefore $\hat{A}_c \subseteq \overline{A_d}^{\sigma(\mathcal{B}_0'^*, \mathcal{B}_0')}$.

(c) From the representation (16.1) of the bidual we now obtain

$$\mathcal{B}_0'' = \bigcup_{c \in (0,\infty)^{\mathbb{N}_0}} \overline{A_c}^{\sigma(\mathcal{B}_0'^*, \mathcal{B}_0')} = \mathcal{B}.$$

$\qquad \square$

Ad 4. We show that $\beta(\mathcal{B}, \mathcal{B}_0')$ is generated by the set of norms $\{p_k ; \ k \in \mathbb{N}_0\}$.

The space $\mathcal{B}_0$ is a Fréchet space, therefore quasi-barrelled, and by Theorem 6.8 the strong topology $\beta(\mathcal{B}, \mathcal{B}_0')$ coincides with the natural topology on $\mathcal{B} = \mathcal{B}_0''$, with neighbourhood base of zero

$$\left\{ \overline{\text{aco}\, U}^{\sigma(\mathcal{B}, \mathcal{B}_0')} ; \ U \in \mathcal{U} \right\},$$

if $\mathcal{U}$ is a neighbourhood base of zero of $\mathcal{B}_0$.

A neighbourhood base of zero of $\mathcal{B}$ is given by

$$\hat{\mathcal{U}} := \left\{ \hat{U}_{m,\varepsilon} ; \ m \in \mathbb{N}_0, \ \varepsilon > 0 \right\},$$

where $\hat{U}_{m,\varepsilon} := \{ f \in \mathcal{B} ;\ p_m(f) \leqslant \varepsilon \}$, and with $U_{m,\varepsilon} := \hat{U}_{m,\varepsilon} \cap \mathcal{B}_0$, the collection

$$\mathcal{U} := \{ U_{m,\varepsilon} ;\ m \in \mathbb{N}_0,\ \varepsilon > 0 \}$$

is a neighbourhood base of zero of $\mathcal{B}_0$.

For $\varepsilon > 0$, $m \in \mathbb{N}_0$ we now show that

$$\overline{U_{m,\varepsilon}}^{\,\sigma(\mathcal{B}, \mathcal{B}_0')} \subseteq \hat{U}_{m,\varepsilon}.$$

For $x \in \mathbb{R}^n$, $|\alpha| \leqslant m$ one has $\partial^\alpha \delta_x \in \mathcal{B}_0'$, $\langle f, \partial^\alpha \delta_x \rangle^* = (-1)^{|\alpha|} \partial^\alpha f(x)$ $(f \in \mathcal{B})$. Therefore

$$\hat{U}_{m,\varepsilon} = \bigcap_{x \in \mathbb{R}^n,\ |\alpha| \leqslant m} \{ f \in \mathcal{B} ;\ |\partial^\alpha f(x)| \leqslant \varepsilon \}$$

is $\sigma(\mathcal{B}, \mathcal{B}_0')$-closed.

On the other hand, if $f \in \hat{U}_{m,\varepsilon}$, then

$$p_m(\eta_k f) \leqslant M_m p_m(\eta_k) p_m(f) \leqslant N_m \varepsilon,$$

where $N_m := M_m \sup_{k \in \mathbb{N}_0} p_m(\eta_k) < \infty$. This implies that

$$\overline{U_{m, N_m \varepsilon}}^{\,\sigma(\mathcal{B}, \mathcal{B}_0')} \supseteq \hat{U}_{m,\varepsilon}.$$

This shows that on $\mathcal{B}$ the neighbourhood bases of zero $\big\{ \overline{U_{m,\varepsilon}}^{\,\sigma(\mathcal{B}, \mathcal{B}_0')} ;\ m \in \mathbb{N}_0,\ \varepsilon > 0 \big\}$ and $\hat{\mathcal{U}}$ are **equivalent**, i.e., they generate the same neighbourhood filter of 0.

Remarks 16.6 (a) If $\Omega \subseteq \mathbb{R}^n$ is open, then $\mathcal{B}_0(\Omega)$ and $\mathcal{B}(\Omega)$ are defined in the same way as above for $\mathbb{R}^n$. For this case it was shown in [DiVo80, Theorem (4.8)] that the bidual of $\mathcal{B}_0(\Omega)$ can be identified in a similar way as above with the space

$$\check{\mathcal{B}}(\Omega) := \{ f \in \mathcal{B}(\Omega) ;\ \text{for every } \alpha \in \mathbb{N}_0^n \text{ there exists an extension } f_\alpha \in C(\overline{\Omega})$$

$$\text{of } \partial^\alpha f \text{ satisfying } f_\alpha|_{\partial \Omega} = 0 \}.$$

(b) Recall from Example 8.4(b) that for bounded Ω the space $\mathcal{B}_0(\Omega) = C_0^\infty(\Omega)$ is reflexive. The bidual $\check{\mathcal{B}}(\Omega)$ indicated in part (a) above, for general open sets $\Omega \subseteq \mathbb{R}^n$, can be considered as a "mixture" of the cases treated in Example 8.4(b) and in the present chapter. △

Notes The example presented in this chapter was first treated by L. Schwartz [Sch54]; see also [Sch66]. The presentation given above follows [DiVo80].

The Krein–Milman Theorem

The Krein–Milman theorem asserts that in a Hausdorff locally convex space all points of a compact convex set can be approximated by convex combinations of its 'corners'. We show that this can be reinforced to the statement that all points of the set are barycentres of probability measures living on the closure of the extreme points of the set. An interesting application to completely monotone functions on $[0, \infty)$ yields Bernstein's theorem concerning Laplace transforms of finite Borel measures on $[0, \infty)$.

Let E be a vector space, $C \subseteq E$. A set $A \subseteq C$ is called an **extreme subset** of C if $A \neq \varnothing$, and if $x, y \in C$, $0 < t < 1$ are such that $(1 - t)x + ty \in A$, then $x, y \in A$. (Convex extreme subsets are also called 'faces'.) An **extreme point** of C is an extreme subset consisting of one point. The set of extreme points of C will be denoted by $\operatorname{ex} C$.

Examples 17.1
(a) As a first example we consider a triangle C in the plane. The extreme points are the vertices. As extreme subsets one obtains, besides the vertices, the edges and the whole triangle.

(b) For the closed unit ball in the plane (or in $\mathbb{R}^n$), the extreme points are the points of the unit sphere.

(c) Looking at the convex hull C of the points $(1, 0, 1)$, $(1, 0, -1)$ and the circle $\{(x, y, 0) \in \mathbb{R}^3 \,;\, x^2 + y^2 = 1\}$ in $\mathbb{R}^3$, the set of extreme points is

$$\{(1, 0, 1), (1, 0, -1)\} \cup \{(x, y, 0)\,;\, (x, y) \in \mathbb{R}^2 \setminus \{(1, 0)\},\ x^2 + y^2 = 1\}.$$

(The point $(1, 0, 0)$ is not extreme as it belongs to the line connecting the points $(1, 0, 1)$ and $(1, 0, -1)$.) This example makes it clear that the set of extreme points of a compact convex set need not be closed.

(d) In Proposition 17.9 it will be shown that the extreme points of the set of probability measures on a Hausdorff compact space are Dirac measures. $\triangle$

© Springer Nature Switzerland AG 2020
J. Voigt, *A Course on Topological Vector Spaces*, Compact Textbooks in Mathematics,
https://doi.org/10.1007/978-3-030-32945-7_17

Remarks 17.2 Let E be a vector space, $\varnothing \neq C \subseteq E$.

(a) If $A_0 \subseteq A_1 \subseteq C$, A_1 is extreme in C, and A_0 is extreme in A_1, then it is easy to see that A_0 is extreme in C.

(b) If $\mathbb{K} = \mathbb{R}$, and $x^* \in E^*$ is bounded below on C, and

$$A := \left\{ x \in C ; \; x^*(x) = \inf_{y \in C} x^*(y) \right\} \neq \varnothing,$$

then it is easy to see that A is an extreme subset of C. △

Lemma 17.3 *Let E be a Hausdorff locally convex space, and let $C \subseteq E$ be non-empty and compact. Then C possesses an extreme point.*

Proof

Without loss of generality, take $\mathbb{K} = \mathbb{R}$. The set

$$\mathcal{A} := \left\{ A \subseteq C ; \; A \text{ closed, extreme in } C \right\}$$

is ordered by inclusion and non-empty (because $C \in \mathcal{A}$). Let $\mathcal{A}_1$ be a chain in $\mathcal{A}$. Then $A_1 := \bigcap \mathcal{A}_1 \neq \varnothing$, because C is compact. It is easy to see that A_1 is extreme, and therefore A_1 is a lower bound of $\mathcal{A}_1$. By Zorn's lemma, there exists a minimal element A_0 of $\mathcal{A}$. Now it will be sufficient to show that A_0 is a singleton.

Assume that there exist $x_1, x_2 \in A_0$, $x_1 \neq x_2$. By Theorem 2.6, there exists $x' \in E'$ such that $x'(x_1) < x'(x_2)$. Then the set

$$A := \left\{ x \in A_0 ; \; x'(x) = \inf_{y \in A_0} x'(y) \right\}$$

is non-empty, because A_0 is compact and x' is continuous. From Remark 17.2 it follows that A is extreme in C. Since of $A \subset A_0$, we obtain a contradiction to the minimality of A_0. □

Theorem 17.4 (Krein–Milman)
Let E be a Hausdorff locally convex space, and let $C \subseteq E$ be compact. Then $C \subseteq \overline{\mathrm{co}}\,\mathrm{ex}\,C$. If C is convex, then $C = \overline{\mathrm{co}}\,\mathrm{ex}\,C$.

Proof

Without loss of generality, take $\mathbb{K} = \mathbb{R}$. Assume that there exists $x_0 \in C \setminus \overline{\mathrm{co}}\,\mathrm{ex}\,C$. Then, by Theorem 2.6, there exists $x' \in E'$ such that

$$x'(x_0) < \inf \left\{ x'(y) ; \; y \in \overline{\mathrm{co}}\,\mathrm{ex}\,C \right\}.$$

By Remark 17.2(b),

$$C_0 := \left\{ x \in C ; \; x'(x) = \inf_{y \in C} x'(y) \right\}$$

is an extreme subset of C, and $C_0 \cap \operatorname{ex} C = \varnothing$. From Lemma 17.3 we know that C_0 possesses an extreme point, which by Remark 17.2(a) is also an extreme point of C; a contradiction.

If additionally C is convex, then $C \subseteq \overline{\operatorname{co}} \operatorname{ex} C \subseteq \overline{\operatorname{co}} C = C$. $\qquad\square$

Remarks 17.5 (a) If E is a normed space, then the closed unit ball $C := B_{E'}$ is $\sigma(E', E)$-compact, by the Banach–Alaoglu theorem. The Krein–Milman theorem implies that $C = \overline{\operatorname{co} \operatorname{ex} C}^{\sigma(E',E)}$. Therefore, if in some Banach space the closed unit ball has too few extreme points, one concludes that this space cannot be the dual of a normed space.

(b) For the space $C([0, 1]; \mathbb{R})$ it is not difficult to see that the only extreme points of the closed unit ball are the constant functions $\mathbf{1}$ and $-\mathbf{1}$. This implies that $C([0, 1]; \mathbb{R})$ is not a dual space.

(c) It is not difficult to show that the closed unit balls of c_0 and $L_1(0, 1)$ have no extreme points. Again, this implies that c_0 and $L_1(0, 1)$ are not dual spaces. $\qquad\triangle$

Remarks 17.6 (a) Rudin [Rud91, Theorem 3.23] proves the following version of the Krein–Milman theorem (without local convexity): Let (E, τ) be a topological vector space with the property that E' separates the points of E, and let $C \subseteq E$ be a compact convex set. Then $C = \overline{\operatorname{co}} \operatorname{ex} C$.

We derive this from Theorem 17.4. Note that the hypotheses imply that $\sigma(E, E') \subseteq \tau$ is a Hausdorff locally convex topology on E. This implies that C is $\sigma(E, E')$-compact, and in fact that $\tau \cap C = \sigma(E, E') \cap C$ (by Lemma 4.12). Therefore Theorem 17.4 shows that

$$C = \overline{\operatorname{co} \operatorname{ex} C}^{\sigma(E,E')} = \overline{\operatorname{co}} \operatorname{ex} C.$$

(b) We recall that in Theorem 17.4 the closed convex hull of the set C need not be compact; see Example 4.10. In contrast to the formulation in Theorem 17.4, in part (a) above the set C is required to be convex.

(c) If E is a non-locally convex topological vector space, then there may exist compact convex sets without extreme points; see [Rob76, Rob77, KaPe80]. $\qquad\triangle$

The Krein–Milman theorem amounts to the statement that every element of C can be approximated by convex combinations of extreme points of C. Next, we modify this statement to the effect that every point of C can be obtained as the barycentre of a probability measure on $\overline{\operatorname{ex}} C$.

Let E be a Hausdorff locally convex space, let $C \subseteq E$ be compact, and let μ be a probability Borel measure on C. Then

$$\int_C x \, d\mu(x) \in E'^*$$

(defined in Chapter 14) is called the **barycentre** of μ.

Corollary 17.7 *Let E be a Hausdorff locally convex space, and let $C \subseteq E$ be a compact set. Then every point of $\overline{\mathrm{co}}\, C$ is the barycentre of a probability Borel measure on $\overline{\mathrm{ex}}\, C$.*

Proof

Without restriction $\mathbb{K} = \mathbb{R}$. The set $C_e := \overline{\mathrm{ex}}\, C$ is compact. Let $x_0 \in \overline{\mathrm{co}}\, C$.

We define the subspace $L \subseteq C(C_e)$ by

$$L := \big\{ x'|_{C_e} \,;\, x' \in E' \big\}.$$

On L we define the linear functional $\varphi : L \to \mathbb{R}$ by

$$\varphi(x'|_{C_e}) := x'(x_0).$$

In order to see that φ is well-defined we note that

$$x'(\overline{\mathrm{co}}\, C) \subseteq x'(\overline{\mathrm{co}}\, C_e) \subseteq \overline{x'(\mathrm{co}\, C_e)} = \Big[\min_{x \in C_e} x'(x), \max_{x \in C_e} x'(x) \Big]. \tag{17.1}$$

Therefore, if $x', y' \in E'$ coincide on C_e, then $x'(x_0) - y'(x_0) = 0$. Moreover, for the sublinear functional $p : C(C_e) \to \mathbb{R}$ given by

$$p(f) := \max_{x \in C_e} f(x) \quad (f \in C(C_e)),$$

(17.1) shows that $\varphi(f) \leqslant p(f)$ for all $f \in L$. Using Theorem A.2 and then Example A.5, we deduce that there exists a probability Borel measure μ on C_e such that $x'(x_0) = \int_{C_e} x'(x)\, d\mu(x)$ for all $x' \in E'$. □

Remarks 17.8 (a) We note that Corollary 17.7 can also be deduced from Proposition 14.4. Indeed, from Theorem 17.4 we know that $C \subseteq \overline{\mathrm{co}}\, C_e$ (with $C_e = \overline{\mathrm{ex}}\, C$). Therefore Proposition 14.4 implies that

$$\overline{\mathrm{co}}\, C \subseteq \overline{\mathrm{co}\, C_e}^{\sigma(E'^*, E')} = \Big\{ \int_{C_e} x\, d\mu(x) \,;\, \mu \in \mathcal{M}_1(C_e) \Big\}.$$

(b) Note that our hypotheses do not imply that $\overline{\mathrm{co}}\, C$ is compact.

(c) Corollary 17.7 can be considered as a starting point of the Choquet theory. One aim of this theory is to investigate the question under what conditions the measure representing a point in $\overline{\mathrm{co}}\, C$ is carried by $\mathrm{ex}\, C$ instead of its closure. We refer to [Phe01] for further information. △

We illustrate Corollary 17.7 with the set of probability measures on a Hausdorff compact space X. We equip $C(X)' = \mathcal{M}(X)$ – see Remark 14.3 – with the vague topology $\tau_v = \sigma(\mathcal{M}(X), C(X))$. Then $\mathcal{M}_1(X)$, the set of regular Borel probability measures, is a (vaguely) compact subset of $\mathcal{M}(X)$, by the Banach–Alaoglu theorem.

Proposition 17.9 *Under the previous hypotheses, the extreme points of $\mathcal{M}_1(X)$ are given by the set $\{\delta_x \,;\, x \in X\}$ of Dirac measures.*

Proof
Let $x \in X$. If $\mu_0, \mu_1 \in \mathcal{M}_1(X)$, $0 < t < 1$ are such that $\delta_x = (1 - t)\mu_0 + t\mu_1$, then $1 = \delta_x(\{x\}) = (1 - t)\mu_0(\{x\}) + t\mu_1(\{x\})$, and this forces $\mu_0(\{x\}) = \mu_1(\{x\}) = 1$, i.e., $\mu_0 = \mu_1 = \delta_x$.

On the other hand, if $\mu \in \mathcal{M}_1(X)$ is not a Dirac measure, then there exists a Borel set $A \subseteq X$ such that $0 < \mu(A) < 1$, and defining

$$\mu_0(B) := \tfrac{1}{\mu(A)}\mu(A \cap B), \quad \mu_1(B) := \tfrac{1}{\mu(X \setminus A)}\mu((X \setminus A) \cap B) \qquad (B \subseteq X \text{ Borel set})$$

one obtains $\mu_0, \mu_1 \in \mathcal{M}_1(X)$, $\mu = \mu(A)\mu_0 + \mu(X \setminus A)\mu_1$. This shows that μ is not an extreme point of $\mathcal{M}_1(X)$. $\qquad\square$

The mapping $X \ni x \mapsto \delta_x \in (\mathrm{ex}\, \mathcal{M}_1(X), \tau_v)$ is continuous – indeed, the mapping $X \ni x \mapsto \langle f, \delta_x \rangle = f(x)$ is continuous for all $f \in C(X)$ –, and Lemma 4.12 implies that this mapping is a homeomorphism. Identifying X and $\mathrm{ex}\, \mathcal{M}_1(X)$ by this homeomorphism, the integral representation of $\mu \in \mathcal{M}_1(X)$ in Corollary 17.7 is given by

$$\mu = \int_X \delta_x \, \mathrm{d}\mu(x);$$

indeed,

$$\langle f, \mu \rangle = \int_X f \, \mathrm{d}\mu = \int_X f(x) \, \mathrm{d}\mu(x) = \int_X \langle f, \delta_x \rangle \, \mathrm{d}\mu(x)$$

for all $f \in C(X)$.

In the context of probability measures on Hausdorff compact spaces we also mention an interesting application of the Krein–Milman theorem in ergodic theory, yielding the existence of ergodic measures for topological dynamical systems; see [EFHN15, Proposition 10.4].

Quite clearly, the Krein–Milman theorem has a strong geometric flavour. In order to illustrate its analytic importance we will treat an application of the Krein–Milman theorem, in particular of Corollary 17.7, to completely monotone functions. The aim is to present a proof of Bernstein's theorem, Theorem 17.12.

Let $E := \mathbb{R}^{[0,\infty)}$, equipped with the product topology τ_s. For $a > 0$ we define the linear mapping $\Delta_a \colon E \to E$,

$$\Delta_a f := f(\cdot + a) - f \qquad (f \in E).$$

A function $f \in E$ is **completely monotone**, if

$$(-1)^n \Delta_{a_n} \cdots \Delta_{a_1} f \geqslant 0$$

for all $n \in \mathbb{N}_0, a_1, \ldots, a_n > 0$.

Remark 17.10 If f is completely monotone, then $f \geqslant 0$ (from the condition for $n = 0$), f is monotone decreasing (from the condition for $n = 1$), and f is convex (from the condition for $n = 2$). (To make the picture more complete: The condition for $n = 2$ implies the 'midpoint convexity' of f, but this together with the monotonicity implies the convexity.) The convexity of f implies that f is continuous on $(0, \infty)$. △

In the following we want to apply Corollary 17.7 to the set

$$C := \big\{ f \in E ; \ f \text{ completely monotone}, \ f(0) = 1 \big\}.$$

It is easy to see that C is a compact convex subset of E.

For $\alpha \in [0, \infty]$ (the one-point compactification of $[0, \infty)$) we define

$$g_\alpha(t) := \begin{cases} e^{-\alpha t}, & \text{if } \alpha \in [0, \infty), \ t \geqslant 0, \\ \mathbf{1}_{\{0\}}(t), & \text{if } \alpha = \infty, \ t \geqslant 0. \end{cases}$$

Then $g_\alpha \in C$ for all $\alpha \in [0, \infty]$; it is elementary to verify the conditions defining complete monotonicity for the functions g_α.

Clearly, the mapping $[0, \infty] \ni \alpha \to g_\alpha \in E$ is continuous, and because it is injective, it is a homeomorphism onto it range.

Lemma 17.11 $\operatorname{ex} C = \big\{ g_\alpha ; \ \alpha \in [0, \infty] \big\}.$

Proof
It is obvious that $g_0, g_\infty \in \operatorname{ex} C$. Clearly, g_0 and g_∞ cannot be the only extreme points of C, because the convex hull of $\{g_0, g_\infty\}$ is not dense in C, which would contradict the Krein–Milman theorem.

Let $g \in \operatorname{ex} C$, $g \neq g_0, g_\infty$. Then there exists $a > 0$ such that $0 < g(a) < 1$. We define

$$f_1 := \frac{1}{g(a)} g(\cdot + a), \quad f_2 := \frac{1}{1 - g(a)} \big(g - g(\cdot + a) \big) = \frac{1}{1 - g(a)} \big(- \Delta_a g \big).$$

Then $f_1, f_2 \in C$ and $g = g(a) f_1 + (1 - g(a)) f_2$. As $g \in \operatorname{ex} C$, this implies that $g = f_1$,

$$g(t) g(a) = g(t + a) \tag{17.2}$$

for all $t > 0$. This equality implies that $g(t) \neq 0$ for all $t > 0$, and as a consequence one deduces that (17.2) holds for all $a, t > 0$. It is well-known that the functional equation (17.2), together with $g(a) < 1$ and the continuity of g on $(0, \infty)$, implies that there exists $\alpha > 0$ such that $g = g_\alpha$.

For $b > 0$ we define the linear mapping $J_b \colon E \to E$, $f \mapsto f(b \cdot)$. Then $J_b \colon C \to C$ is bijective, and a function $g \in C$ is an extreme point if and only if $J_b g = g(b \cdot)$ is an extreme point. Note that $J_b g_\alpha = g_\alpha (b \cdot) = g_{b\alpha}$ for all $\alpha, b > 0$. Since the functions g_α, for $\alpha > 0$, are the only possible extreme points besides g_0, g_∞, it follows that all of them belong to $\mathrm{ex}\, C$. □

Theorem 17.12 (Bernstein)

(a) Let $f \in E = \mathbb{R}^{[0,\infty)}$ be completely monotone, $f(0) = 1$. Then there exists a unique probability Borel measure μ on $[0, \infty]$ such that

$$f(t) := \int_{[0,\infty]} g_\alpha(t) \, d\mu(\alpha) = \int_{[0,\infty)} e^{-\alpha t} \, d\mu(\alpha) + \mu(\{\infty\})\mathbf{1}_{\{0\}} \qquad (17.3)$$

for all $t \geqslant 0$.

Moreover, f is infinitely differentiable on $(0, \infty)$, and f is continuous (at 0) if and only if $\mu(\{\infty\}) = 0$.

(b) Conversely, if μ is a probability Borel measure μ on $[0, \infty]$, then (17.3) defines a completely monotone function satisfying $f(0) = 1$.

Proof

(a) From Corollary 17.7 we know that there exists a probability Borel measure ν on $C_e = \overline{\mathrm{ex}}\, C$ such that $f = \int_{C_e} g \, d\nu(g)$. From Lemma 17.11 we know that $\mathrm{ex}\, C$ is closed and in fact homeomorphic to $[0, \infty]$. This implies that one can transport ν to $[0, \infty]$ and thereby obtain a probability Borel measure μ on $[0, \infty]$ such that $f = \int_{[0,\infty]} g_\alpha \, d\mu(\alpha)$. Then the continuity of the evaluation functionals $E \ni f \mapsto f(t) \in \mathbb{R}$ implies (17.3) ($t \in [0, \infty)$).

In order to show the uniqueness of μ we define the functions $\psi_t \in C[0, \infty]$ for $t \in [0, \infty)$,

$$\psi_t(\alpha) := e^{-\alpha t} \quad (\alpha \in [0, \infty])$$

(where $\psi_t(\infty) = 0$, for $t > 0$, $\psi_0(\infty) = 1$). Then $\mathrm{lin}\{\psi_t \, ; \, t \in [0, \infty)\} \subseteq C[0, \infty]$ is an algebra containing the constant functions and separating the points of $[0, \infty]$, so it is dense in $C[0, \infty]$, by the Stone–Weierstraß theorem (see [Bou74, X, § 4.2, Théorème 3]). It follows that the measure μ, identified with the corresponding functional $\mu \in C[0, \infty]'$, is uniquely determined by its values

$$\mu(\psi_t) = \int_{[0,\infty]} e^{-\alpha t} \, d\mu(\alpha) = f(t) \quad (t \in [0, \infty)).$$

By the uniqueness in the Riesz–Markov theorem, this implies that the measure μ is uniquely determined.

From the representation (17.3) one deduces, by differentiation under the integral sign, that f is infinitely differentiable. The number $\mu(\{\infty\})$ corresponds to the jump of f at 0.

(b) As g_α is completely monotone for all $\alpha \in [0, \infty]$, the formula (17.3) implies that f is completely monotone. From $\mu([0, \infty]) = 1$ one obtains $f(0) = \int_{[0,\infty]} 1 \, d\mu(\alpha) = 1$.   □

Remarks 17.13 (a) Bernstein's theorem implies that the set of continuous completely monotone functions $f \colon [0, \infty) \to \mathbb{R}$ is equal to the set of Laplace transforms of finite (positive) Borel measures on $[0, \infty)$.

(b) Theorem 17.12 also shows that completely monotone functions are infinitely differentiable on $(0, \infty)$, and satisfy $f(0) \geqslant \limsup_{t \to 0+} f(t)$ and

$$(-1)^n f^{(n)} \geqslant 0 \quad (n \in \mathbb{N}_0), \tag{17.4}$$

where the latter inequalities are a consequence of (17.3). In fact, these properties characterise complete monotonicity. The sufficiency of (17.4) is obtained by a (careful!) repeated use of the mean value theorem.   △

Notes The Krein–Milman theorem was first shown in [KrMi40], for duals of Banach spaces E and the topology $\sigma(E', E)$. The author could not find an original source for Corollary 17.7; however, the idea to represent points of the closed convex hull of a bounded set S as 'barycentres' of suitable positive functionals on the set of continuous functions on S appears already in the paper of Krein and Šmulian [KrŠm40, Theorem 2]. Bernstein's theorem is a classical example for illustrating the Krein–Milman theorem in the form of Corollary 17.7. Our exposition is based on [Phe01, Sec. 2], [LMNS10, Sec. 14.7].

The Hahn–Banach Theorem

For a vector space E over $\mathbb{C}$, we denote by E_0 the **associated vector space over** $\mathbb{R}$, i.e., the vector space obtained by restricting the scalar multiplication to $\mathbb{R} \times E$.

Lemma A.1 *Let E be a vector space over $\mathbb{C}$. The mapping*

$$j \colon (E^*)_0 \to (E_0)^*, \quad \varphi \to \operatorname{Re}\varphi,$$

is an $\mathbb{R}$-linear isomorphism. For $\varphi \in E^*$ one has

$$\varphi(x) = \operatorname{Re}\varphi(x) - \mathrm{i}\operatorname{Re}\varphi(\mathrm{i}x) \qquad (x \in E).$$

Proof
It is easy to check that j is an $\mathbb{R}$-linear mapping from $(E^*)_0$ to $(E_0)^*$.

The isomorphism property is proved by checking that the inverse of j is given by

$$j^{-1}(\psi)(x) = \psi(x) - \mathrm{i}\psi(\mathrm{i}x) \qquad (x \in E, \ \psi \in (E_0)^*). \qquad \square$$

Theorem A.2 (Hahn–Banach, analytic form)
Let E be a vector space, and let $p \colon E \to \mathbb{R}$ sublinear. Let $L \subseteq E$ be a subspace, and let $\varphi \in L^*$ satisfying

$$\operatorname{Re}\varphi(x) \leqslant p(x) \qquad (x \in L).$$

Then there exists an extension $\Phi \in E^*$ of φ such that

$$\operatorname{Re}\Phi(x) \leqslant p(x) \qquad (x \in E).$$

© Springer Nature Switzerland AG 2020
J. Voigt, *A Course on Topological Vector Spaces*, Compact Textbooks in Mathematics,
https://doi.org/10.1007/978-3-030-32945-7

Proof

Because of Lemma A.1 it is sufficient to treat the case of a real vector space. Thus, assume that $\mathbb{K} = \mathbb{R}$.

(i) In the first step we show that there exist maximal extensions of φ. In order to do this we introduce the set

$$\mathcal{Z} := \big\{ \psi ; \ \psi \text{ linear extension of } \varphi \text{ to a subspace } \operatorname{dom} \psi \subseteq E, $$

$$\psi(x) \leqslant p(x) \ (x \in \operatorname{dom} \psi) \big\}.$$

(Here, $\operatorname{dom} \psi$ denotes the domain of ψ, the extension property of ψ implies in particular that $\operatorname{dom} \psi \supseteq L$.) Then obviously the set $\mathcal{Z}$ is ordered by the inclusion of the graphs of its elements. Let $\mathcal{Y} \subseteq \mathcal{Z}$ be a chain (a linearly ordered subset). Then an upper bound $\eta \in \mathcal{Z}$ of $\mathcal{Y}$ is obtained by

$$\operatorname{dom} \eta := \bigcup_{\psi \in \mathcal{Y}} \operatorname{dom} \psi, \quad \eta(x) := \psi(x) \ (\psi \in \mathcal{Y}, \ x \in \operatorname{dom} \psi).$$

Now Zorn's lemma implies that $\mathcal{Z}$ contains maximal elements (with respect to the order defined on $\mathcal{Z}$).

(ii) To show the assertion it now is sufficient to show that maximal elements $\psi \in \mathcal{Z}$ satisfy $\operatorname{dom} \psi = E$.

Let $\psi \in \mathcal{Z}$, $F := \operatorname{dom} \psi$, and let $a \in E \setminus F$. Then the elements $z \in \operatorname{lin}(F \cup \{a\})$ are given by $z = x + \lambda a$, with $x \in F$ and $\lambda \in \mathbb{R}$ uniquely determined by z, and the extensions of ψ to $\operatorname{lin}(F \cup \{a\})$ are given by

$$\psi_\gamma(x + \lambda a) = \psi(x) + \lambda \gamma \quad (x \in F, \ \lambda \in \mathbb{R}),$$

with $\gamma = \psi_\gamma(a) \in \mathbb{R}$. We have to show that one can choose γ such that

$$\psi_\gamma(x + \lambda a) \leqslant p(x + \lambda a) \quad (x \in F, \ \lambda \in \mathbb{R}).$$

It is easy to see that it is sufficient to have the last inequality for $\lambda = \pm 1$. So, we have to show that there exists $\gamma \in \mathbb{R}$ such that

$$\begin{aligned} \psi(x) + \gamma &\leqslant p(x + a) \quad (x \in F), \\ \psi(y) - \gamma &\leqslant p(y - a) \quad (y \in F); \end{aligned} \tag{A.1}$$

or, put together,

$$\psi(y) - p(y - a) \leqslant \gamma \leqslant p(x + a) - \psi(x) \quad (x, y \in F).$$

Now, for all $x, y \in F$ one has

$$\psi(y) + \psi(x) = \psi(y + x) \leqslant p(y + x) \leqslant p(y - a) + p(x + a),$$
$$\psi(y) - p(y - a) \leqslant p(x + a) - \psi(x);$$

hence

$$S := \sup_{y \in F} \big(\psi(y) - p(y - a)\big) \leqslant \inf_{x \in F} \big(p(x + a) - \psi(x)\big) =: I.$$

As a consequence, the desired inequalities (A.1) are satisfied for all $\gamma \in [S, I]$; hence ψ is not maximal. $\qquad\square$

Corollary A.3 *Let (E, p) be a semi-normed space. Let $L \subseteq E$ be a subspace, and let $\varphi \in L^*$ be such that*

$$|\varphi(x)| \leqslant p(x) \quad (x \in L).$$

Then there exists an extension $\Phi \in E^*$ of φ such that

$$|\Phi(x)| \leqslant p(x) \quad (x \in E).$$

Proof

As a semi-norm, p is a sublinear functional. By Theorem A.2 there exists an extension $\Phi \in E^*$ of φ satisfying

$$\operatorname{Re} \Phi(x) \leqslant p(x) \quad (x \in E).$$

Let $x \in E$. Then there exists $\gamma \in \mathbb{K}$ with $|\gamma| = 1$ such that $|\Phi(x)| = \gamma \Phi(x) = \Phi(\gamma x)$. This implies

$$|\Phi(x)| = \operatorname{Re} \Phi(\gamma x) \leqslant p(\gamma x) = p(x). \qquad\square$$

Corollary A.4 *Let (E, p) be a semi-normed space, and let $x \in E$ be such that $p(x) \neq 0$. Then there exists $x' \in E'$ such that $\|x'\| = 1$ and $\langle x, x' \rangle = p(x)$.*

Proof

Apply Corollary A.3 to $L := \operatorname{lin}\{x\}$ and $\varphi \in L^*$, defined by $\varphi(\lambda x) := \lambda p(x)$ $(\lambda \in \mathbb{K})$. $\qquad\square$

We include the following example to illustrate the usefulness of the "sublinear functional version" (Theorem A.2) of the Hahn–Banach theorem.

Example A.5

Let K be a compact topological space. On $C(K)$ we define $p \colon C(K) \to \mathbb{R}$,

$$p(f) := \max_{x \in K} \operatorname{Re} f(x) \quad (f \in C(K)).$$

It is easy to see that then p is a sublinear functional.

Let $\varphi \in C(K)^*$ with

$$\operatorname{Re} \varphi(f) \leqslant p(f) \quad (f \in C(K)). \tag{A.2}$$

If $f \in C(K)$ with $\operatorname{Re} f(x) = 0$ for all $x \in K$, then one deduces that $\pm \operatorname{Re} \varphi(f) \leqslant 0$, i.e., $\varphi(f)$ is purely imaginary. This implies that $\varphi(f) \in \mathbb{R}$ for all real valued $f \in C(K)$. Moreover, if $0 \leqslant f \in C(K)$, then $-\varphi(f) = \varphi(-f) \leqslant p(-f) \leqslant 0$, i.e., φ is a positive linear functional. Finally, from

$$\varphi(\pm\mathbf{1}) \leqslant p(\pm\mathbf{1}) = \pm 1$$

we obtain $\varphi(\mathbf{1}) = 1$. This means that in the representation from the Riesz–Markov theorem (see [Rud87, Theorem 2.14]) the functional φ corresponds to a probability measure.

On the other hand, if μ is a Borel probability measure on K, then clearly the functional φ defined by

$$\varphi(f) := \int_K f \, \mathrm{d}\mu \quad (f \in C(K))$$

satifies (A.2).   △

Notes For the case of real scalars, Theorem A.2 was proved by Hahn [Hah27, Satz III], with a norm instead of a sublinear functional, and by Banach [Ban29, Théorème 1 and its proof], [Ban32, II, § 2, Théorème 1] with general sublinear functionals. The extension (in the form of Corollary A.3) to the case of complex scalars is due to Bohnenblust and Sobczyk [BoSo38, Theorem 1].

Baire's Theorem and the Uniform Boundedness Theorem

Theorem B.1 (Baire)

Let (X, d) be a complete semi-metric space, and let $(U_n)_{n \in \mathbb{N}}$ be a sequence of dense open subsets of X. Then $\bigcap_{n \in \mathbb{N}} U_n$ is dense in X.

Proof

Let $x_0 \in X$, $r_0 > 0$. It is sufficient to show that then $B[x_0, r_0] \cap \bigcap_{n \in \mathbb{N}} U_n \neq \varnothing$.

We claim that there exist a sequence (x_n) in X and a null sequence (r_n) in $(0, \infty)$ such that $B[x_n, r_n] \subseteq U_n \cap B(x_{n-1}, r_{n-1})$ $(n \in \mathbb{N})$. Indeed, assume that $x_1, \dots, x_{n-1}$ and $r_1, \dots, r_{n-1}$ are chosen. Then there exist $x_n \in U_n \cap B(x_{n-1}, r_{n-1})$ and $r_n \in (0, r_{n-1}/2)$ such that $B[x_n, r_n] \subseteq U_n \cap B(x_{n-1}, r_{n-1})$.

Now the completeness of X implies that

$$\varnothing \neq \bigcap_{n \in \mathbb{N}} B[x_n r_n] \subseteq B[x_0, r_0] \cap \bigcap_{n \in \mathbb{N}} U_n. \qquad \square$$

A topological space X is called a **Baire space** if for any sequence $(U_n)_n \in \mathbb{N}$ of dense open subsets of X the intersection $\bigcap_{n \in \mathbb{N}} U_n$ is dense in X. With this terminology, Theorem B.1 states that every complete semi-metric space is a Baire space.

The following notions are used in Chapter 7. In a topological space X, a set $A \subseteq X$ is called **meagre** (or of **first category**) if A is contained in the union of a sequence of closed sets with empty interior. A set B is called **residual** (or **comeagre**) if $X \setminus B$ is meagre, i.e., if B is the intersection of a sequence of sets with dense interior.

A G_δ-**set** in X is the intersection of a sequence of open sets, whereas an F_σ-**set** is the union of a sequence of closed sets. (As a consequence, a set is a G_δ-set if and only if its complement is an F_σ-set.) In terms of these notions it follows that dense G_δ-sets are residual.

© Springer Nature Switzerland AG 2020

J. Voigt, *A Course on Topological Vector Spaces*, Compact Textbooks in Mathematics, https://doi.org/10.1007/978-3-030-32945-7

Proposition B.2 *For a topological space X the following properties are equivalent:*

(i)   *X is a Baire space;*

(ii) *every residual set is dense in X;*

(iii) *every meagre set has empty interior.*

In particular, a subset of a non-empty Baire space X cannot be residual and meagre simultaneously.

Proof

The equivalences are obvious. If $A \subseteq X$ would be residual and meagre, then $X = (X \setminus A) \cup A$ would be meagre.   □

A very important consequence of these notions is the uniform boundedness theorem; we will present only the version for linear functionals, to which we refer at various places in the main text. The reader should keep in mind that, in view of Baire's theorem, it applies to Banach spaces.

Theorem B.3 (Uniform boundedness theorem)

Let (E, p) be a semi-normed space, assume that E is a Baire space, and let $B \subseteq E'$ be $\sigma(E', E)$-bounded. Then $\sup_{x' \in B} \|x'\| < \infty$.

Proof

(i) We start with a preliminary statement. Let $x' \in E'$, $x \in E$, $r > 0$. Then

$$r\|x'\| \leqslant \sup_{y \in B(x,r)} |x'(y)|.$$

Indeed, for $z \in E$, $\|z\| < r$, one has

$$|x'(z)| \leqslant 1/2\big(|x'(x+z)| + |x'(x-z)|\big) \leqslant \sup_{y \in B(x,r)} |x'(y)|.$$

(ii) For $n \in \mathbb{N}$, we define $A_n := \big\{x \in E;\ \sup_{x' \in B} |x'(x)| \leqslant n\big\}$. Then $A_n = \bigcap_{x' \in B} x'^{-1}(B_{\mathbb{K}}[0, n])$ is a closed subset of E, and $\bigcup_{n \in \mathbb{N}} A_n = E$. Then Theorem B.1 implies that there exists $n \in \mathbb{N}$ such that $\mathring{A}_n \neq \varnothing$; see Proposition B.2. This implies that there exist $x \in A_n$, $r > 0$ such that $B(x, r) \subseteq A_n$. Then part (i) of the proof implies

$$\sup_{x' \in B} \|x'\| \leqslant \frac{n}{r}.$$ □

Remarks B.4 (a) The adjective 'uniform' refers to the fact that the norm of the functionals is the supremum over the vectors of the closed unit ball; so the boundedness is uniform with respect to $x \in B[0, 1]$.

(b) We have stated the theorem only for functionals; the proof for bounded linear operators mapping E to a normed space is analogous.   △

Notes Baire proved Theorem B.1 in his Thèse [Bai99, II, 59] for the case of the real line. (Incidentally, in that paper he already introduced the notions of sets of 'première catégorie' and 'deuxième catégorie'.) The first versions of the uniform boundedness theorem were proved by Hahn [Hah22, §2] for linear functionals and by Banach [Ban22, II, Théorème 5] for linear operators. Neither of these references resort to Baire's category theorem, Theorem B.1. The uniform boundedness theorem, Theorem B.3, we have presented can be proved in much more generality; we refer to [Rud91, Theorems 2.5 (Banach–Steinhaus) and 2.6].

References

[ABHN11] W. Arendt, C. Batty, M. Hieber, and F. Neubrander: *Vector-valued Laplace Transforms and Cauchy Problems*. Second edition. Birkhäuser, Basel, 2011.

[Ala40] L. Alaoglu: Weak topologies of normed linear spaces. *Ann. of Math.* (2) **41**, 252–267 (1940).

[Are47] R. Arens: Duality in linear spaces. *Duke Math. J.* **14**, 787–794 (1947).

[ArNi00] W. Arendt and N. Nikolski: Vector-valued holomorphic functions revisited. *Math. Z.* **234**, 777–805 (2000).

[Bai99] R. Baire: *Sur les fonctions de variables réelles*. Annali di Matematica, Imprimerie Bernardoni de C. Rebeschini & C., Milan, 1899.

[Ban22] S. Banach: Sur les opérations dans les ensembles abstraits et leurs applications aux équations intégrales. *Fund. Math.* **3**, 133–181 (1922).

[Ban29] S. Banach: Sur les fonctionnelles linéaires II. *Studia Math.* **1**, 223–239 (1929).

[Ban32] S. Banach: *Théorie des Opérations Linéaires*. Chelsea, New York, 1932.

[Bar85] J. A. Barroso: *Introduction to Holomorphy*. North-Holland, Amsterdam, 1985.

[Bau90] H. Bauer: *Maß- und Integrationstheorie*. Walter de Gruyter, Berlin, 1990.

[Bog07] V. I. Bogachev: *Measure Theory*, Volume 1. Springer, Berlin, 2007.

[BoSm17] V. I. Bogachev and O. G. Smolyanov: *Topological Vector Spaces and Their Applications*. Springer, Cham, Switzerland, 2017.

[BoSo38] H. F. Bohnenblust and A. Sobczyk: Extensions of functionals on complex linear spaces. *Bull. Amer. Math. Soc.* **44**, 91–93 (1938).

[Bou07a] N. Bourbaki: *Espaces Vectoriels Topologiques*, Chap. 1 à 5. Réimpression inchangée de l'édition originale de 1981. N. Bourbaki et Springer, Berlin, 2007.

[Bou07b] N. Bourbaki: *Intégration*, Chap. 1 à 4. Réimpression inchangée de l'édition originale de 1965. N. Bourbaki et Springer, Berlin, 2007.

[Bou07c] N. Bourbaki: *Topologie Générale*, Chap. 1 à 4. Réimpression inchangée de l'édition originale de 1971. N. Bourbaki et Springer, Berlin, 2007.

[Bou38] N. Bourbaki: Sur les espaces de Banach. *C. R. Acad. Sci. Paris* **206**, 1701–1704 (1938).

[Bou64a] N. Bourbaki: *Espaces Vectoriels Topologiques*, Chap. I et II. Hermann, Paris, 1964.

[Bou64b] N. Bourbaki: *Espaces Vectoriels Topologiques*, Chap. III à V. Hermann, Paris, 1964.

[Bou74] N. Bourbaki: *Topologie Générale*, Chap. 5 à 10. Hermann, Paris, 1974.

[Con90] J. B. Conway: *A Course in Functional Analysis*. Second edition. Springer, New York, 1990.

[Die40] J. Dieudonné: Topologies faibles dans les espaces vectoriels. *C. R. Acad. Sci. Paris* **211**, 94–97 (1940).

© Springer Nature Switzerland AG 2020
J. Voigt, *A Course on Topological Vector Spaces*, Compact Textbooks in Mathematics,
https://doi.org/10.1007/978-3-030-32945-7

[Die42]   J. Dieudonné: La dualité dans les espaces vectoriels topologiques. *Ann. Sci. École Norm. Sup. (3)* **59**, 107–139 (1942).

[Die78]   P. Dierolf: *Zwei Räume regulärer temperierter Distributionen.* Habilitationsschrift, Fachbereich Mathematik der Universität München. 1978.

[Die84]   J. Diestel: *Sequences and Series in Banach Spaces.* Springer-Verlag, New York, 1984.

[DiSc49]   J. Dieudonné and L. Schwartz: La dualité dans les espaces $(\mathcal{F})$ et $(\mathcal{LF})$. *Ann. Inst. Fourier* **1**, 61–101 (1949).

[DiVo80]   P. Dierolf and J. Voigt: Calculation of the bidual for some function spaces. Integrable distributions. *Math. Ann.* **253**, 63–87 (1980).

[Dun38]   N. Dunford: Uniformity in linear spaces. *Trans. Amer. Math. Soc.* **44**, 305–356 (1938).

[DuPe40]   N. Dunford and B. J. Pettis: Linear operators on summable functions. *Trans. Amer. Math. Soc.* **47**, 323–392 (1940).

[DuSc58]   N. Dunford and J. Schwartz: *Linear Operators, Part I: General Theory.* John Wiley & Sons, New York, 1958.

[Ebe47]   W. F. Eberlein: Weak compactness in Banach spaces. I. *Proc. Nat. Acad. Sci. U. S. A.* **33**, 51–53 (1947).

[Edw65]   R. E. Edwards: *Functional Analysis, Theory and Applications.* Holt, Rinehart and Winston, New York, 1965.

[EFHN15]   T. Eisner, B. Farkas, M. Haase, and R. Nagel: *Operator Theoretic Aspects of Ergodic Theory.* Springer, Cham, Switzerland, 2015.

[Eva98]   L. C. Evans: *Partial Differential Equations.* Americal Mathematical Society, Providence, R.I., 1998.

[Flo80]   K. Floret: *Weakly Compact Sets.* Springer-Verlag, Berlin, 1980.

[Gov80]   W. Govaerts: A productive class of angelic spaces. *J. London Math. Soc. (2)* **22**, 355–364 (1980).

[GrE92]   K.-G. Grosse-Erdmann: *The Borel–Okada theorem revisited.* Habilitationsschrift. Hagen: Fernuniversität Hagen, 1992.

[Gro50]   A. Grothendieck: Sur la complétion du dual d'un espace vectoriel localement convexe. *C. R. Acad. Sci. Paris* **230**, 605–606 (1950).

[Gro52]   A. Grothendieck: Critères de compacité dans les espaces fonctionnels généraux. *Amer. J. Math.* **74**, 168–186 (1952).

[Gro54]   A. Grothendieck: Sur les espaces (F) et (DF). *Summa Brasil. Math.* **3**, 57–123 (1954).

[Gro73]   A. Grothendieck: *Topological Vector Spaces.* Existed previously as mimeographed notes of a course "Espaces Vectoriels Topologiques" from 1954. Gordon and Breach, New York, 1973.

[Hah22]   H. Hahn: Über Folgen linearer Operationen. *Monatsh. Math.* **32**, 3–88 (1922).

[Hah27]   H. Hahn: Über lineare Gleichungssysteme in linearen Räumen. *J. Reine Angew. Math.* **157**, 214–229 (1927).

[Hor66]   J. Horváth: *Topological Vector Spaces and Distributions.* Addison-Wesley, Reading, MA, 1966.

[Jar81]   H. Jarchow: *Locally Convex Spaces.* B. G. Teubner, Stuttgart, 1981.

[Kąk87]   J. Kąkol: Note on compatible vector topologies. *Proc. Amer. Math. Soc.* **99**, 690–692 (1987).

[KaPe80]   N. J. Kalton and N. T. Peck: A re-examination of the Roberts example of a compact convex set without extreme points. *Math. Ann.* **253**, 89–101 (1980).

[Kha82]   S. M. Khaleelulla: *Counterexamples in Topological Vector Spaces.* Springer, Berlin, 1982.

[Kōm64]   Y. Kōmura: Some examples on linear topological spaces. *Math. Ann.* **153**, 150–162 (1964).

[Köt50]   G. Köthe: Über die Vollständigkeit einer Klasse lokalkonvexer Räume. *Math. Z.* **52**, 627–630 (1950).

[Köt66]   G. Köthe: *Topologische Lineare Räume I.* 2nd edition. Springer, Berlin, 1966.

[Kre37] M. Krein: Sur quelques questions de la géometrie des ensembles convexes situés dans un espace linéaire normé et complet. *C. R. Acad. Sci. URSS* **14**, 5–7 (1937).

[KrMi40] M. Krein and D. Milman: On extreme points of regular convex sets. *Studia Math.* **9**, 133–138 (1940).

[KrŠm40] M. Krein and V. Šmulian: On regularly convex sets in the space conjugate to a Banach space. *Ann. Math.* **41**, 556–583 (1940).

[LMNS10] J. Lukeš, J. Malý, I. Netuka, and J. Spurný: *Integral Representation Theory. Applications to Convexity, Banach Spaces and Potential Theory.* W. de Gruyter, Berlin, 2010.

[Mac46] G. W. Mackey: On convex topological linear spaces. *Trans. Amer. Math. Soc.* **60**, 519–537 (1946).

[Maz30] S. Mazur: Über die kleinste konvexe Menge, die eine gegebene kompakte Menge enthält. *Studia Math.* **2**, 7–9 (1930).

[Maz33] S. Mazur: Über konvexe Mengen in linearen normierten Räumen. *Studia Math.* **4**, 70–84 (1933).

[MeVo97] R. Meise and D. Vogt: *Introduction to Functional Analysis.* Clarendon Press, Oxford, 1997.

[Osb14] M. S. Osborne: *Locally Convex Spaces.* Springer, Cham, Switzerland, 2014.

[Phe01] R. R. Phelps: *Lectures on Choquet's Theorem.* 2nd edition. Springer, Berlin, 2001.

[Pry71] J. D. Pryce: A device of R. J. Whitley's applied to pointwise compactness in spaces of continuous functions. *Proc. London Math. Soc. (3)* **23**, 532–546 (1971).

[ReSi80] M. Reed and B. Simon: *Methods of Modern Mathematical Physics, vol. I: Functional Analysis.* Revised and enlarged edition. Academic Press, San Diego, 1980.

[Rob76] J. W. Roberts: Pathological compact convex sets in the spaces L_p, $0 < p < 1$. The Altgeld Book, Functional Analysis Seminar. University of Illinois, 1976.

[Rob77] J. W. Roberts: A compact convex set with no extreme points. *Studia Math.* **60**, 255–266 (1977).

[RoRo73] A. P. Robertson and W. Robertson: *Topological Vector Spaces.* 2nd edition. Cambridge Univ. Press, Cambridge, 1973.

[Rud87] W. Rudin: *Real and Complex Analysis.* 3rd edition. McGraw-Hill, New York, 1987.

[Rud91] W. Rudin: *Functional Analysis.* 2nd edition. McGraw-Hill, New York, 1991.

[Sch54] L. Schwartz: Les distributions sommables. *Séminaire Schwartz* **1**, 1–7 (1953/1954).

[Sch66] L. Schwartz: *Théorie des Distributions.* first published in two volumes: I (1950), II (1951). Hermann, Paris, 1966.

[Sch71] H. H. Schaefer: *Topological Vector Spaces.* 3rd edition. Springer, New York, 1971.

[Seb50] J. Sebastião e Silva: As funções analíticas e a análise funcional. *Port. Math.* **9**, 1–130 (1950).

[Sie28] W. Sierpiński: Sur les ensembles complets d'un espace *(D)*. *Fund. Math.* **11**, 203–205 (1928).

[Sim71] B. Simon: Distributions and their Hermite expansions. *J. Math. Phys.* **12**, 140–148 (1971).

[Šmu40] V. Šmulian: Über lineare topologische Räume. *Rec. Math. [Mat. Sbornik] NS.* **7 (49)**, 425–448 (1940).

[Trè67] F. Trèves: *Topological Vector Spaces, Distributions and Kernels.* Academic Press, New York, 1967.

[Tyc30] A. Tychonoff: Über die topologische Erweiterung von Räumen. *Math. Ann.* **102**, 544–561 (1930).

[Voi92] J. Voigt: On the convex compactness property for the strong operator topology. *Note Mat.* **XII**, 259–269 (1992).

[Wer18] D. Werner: *Funktionalanalysis.* 8th edition. Springer, Berlin, 2018.

[Wil70] S. Willard: *General Topology.* Addison-Wesley, Reading, MA, 1970.

[Wil78] A. Wilansky: *Modern Methods in Topological Vector Spaces.* McGraw-Hill, Inc., New York, 1978.

[Yos80] K. Yosida: *Functional Analysis.* 6th edition. Springer, Berlin, 1980.

Index of Notation

$\mathbb{N}$	set of natural numbers $\mathbb{N} = \{1, 2, \dots\}$
$\mathbb{N}_0$	$\mathbb{N} \cup \{0\}$
$\mathbb{K}$	field of real numbers $\mathbb{R}$ or complex numbers $\mathbb{C}$, page 4
$\mathcal{P}(X)$	power set of a set X, page 1
$\mathrm{top}\,\mathcal{S}$	topology generated by a collection of sets, page 3
$\tau \cap C$	induced topology on a subset C of a topological space (X, τ), page 29
$B_X(x, r),\ B_d(x, r)$	open ball with centre x and radius r, in a semi-metric space (X, d), with omitted subscript 'X' or 'd' if evident from the context, page 2
$B_X[x, r],\ B_d[x, r]$	closed ball as before, page 2
B_E	closed unit ball in a normed space E, page 26
E^*	algebraic dual of the vector space E, vector space of all linear functionals on E, page 6
$E',\ (E, \tau)'$	vector space of all continuous linear functionals on a topological vector space E, page 8
p_A	Minkowski functional (of a convex and absorbing set A), page 11
q_B	semi-norm on E associated with a set $B \subseteq F$, for a dual pair $\langle E, F \rangle$; the Minkowski functional of B°, page 26
$\langle E, F \rangle$	dual pair of vector spaces over the same field, page 6
b_1, b_2	the mappings $b_1 \colon E \to F^*$, $b_2 \colon F \to E^*$, for a dual pair $\langle E, F \rangle$, page 6
$\mathcal{B}_\sigma(F, E)$	collection of $\sigma(F, E)$-bounded subsets of F, in a dual pair $\langle E, F \rangle$, page 26
$\mathcal{E}, \mathcal{C}, \mathcal{B}_\beta, \mathcal{B}_\sigma$	certain subcollections of $\mathcal{B}_\sigma(E', E)$, page 51
$\tau_{\mathcal{M}}$	polar topology on E associated with a collection $\mathcal{M} \subseteq \mathcal{B}_\sigma(F, E)$ in a dual pair $\langle E, F \rangle$, page 26
τ_P	topology generated by a set P of semi-norms, page 7

© Springer Nature Switzerland AG 2020

J. Voigt, *A Course on Topological Vector Spaces*, Compact Textbooks in Mathematics, https://doi.org/10.1007/978-3-030-32945-7

Index

© Springer Nature Switzerland AG 2020
J. Voigt, *A Course on Topological Vector Spaces*, Compact Textbooks in Mathematics,
https://doi.org/10.1007/978-3-030-32945-7

Printed in the United States
By Bookmasters